Math Talks for Undergraduates

Springer
New York
Berlin
Heidelberg
Barcelona
Hong Kong
London
Milan
Paris
Singapore
Tokyo

A First Course in Calculus
1986, ISBN 96201-8

Calculus of Several Variables
1987, ISBN 96405-3

Introduction to Linear Algebra
1986, ISBN 96205-0

Linear Algebra, Third Edition
1987, ISBN 0-387-96412-6

Undergraduate Algebra, Second Edition
1990, ISBN 0-387-97279-X

Undergraduate Analysis, Second Edition
1983, ISBN 0-387-94841-4

Complex Analysis, Fourth Edition
1999, ISBN 0-387-98592-1

BOOKS BY LANG PUBLISHED BY
SPRINGER-VERLAG

CHALLENGES • THE FILE • The Beauty of Doing Mathematics • Math! Encounters with High School Students • Geometry: A High School Course (with Gene Murrow) • Basic Mathematics • First Course in Calculus • Calculus of Several Variables • Introduction to Linear Algebra • Elliptic Curves: Diophantine Analysis • Introduction to Diophantine Approximations • Introduction to Algebraic and Abelian Functions • Abelian Varieties • $SL_2(\mathbf{R})$ • Cyclotomic Fields I and II • Survey of Diophantine Geometry • Number Theory III • Elliptic Functions • Fundamentals of Diophantine Geometry • Modular Units (with Daniel Kubert) • Introduction to Modular Forms • Complex Multiplication • Riemann–Roch Algebra (with William Fulton) • Introduction to Arakelov Theory

Serge Lang

Math Talks for Undergraduates

With 16 Illustrations

Springer

Serge Lang
Department of Mathematics
Yale University
New Haven, CT 06520
USA

Mathematics Subject Classification (1991): 00A05, 00A06

Library of Congress Cataloging in Publication Data
Lang, Serge, 1927–
    Math talks for undergraduates / Serge Lang.
        p.   cm.
    Includes bibliographical references (p.   –   )
    ISBN 0-387-98749-5 (pbk. : alk. paper)
    1. Mathematics.   I. Title.
QA7.L286   1999
510--dc21   98-55410

Printed on acid-free paper.

Production coordinated by Brian Howe and managed by Terry Kornak; manufacturing supervised by Thomas King.
Typeset by TechBooks, Fairfax, VA.
Printed and bound by Edwards Brothers, Inc., Ann Arbor, MI.
Printed in the United States of America.

9 8 7 6 5 4 3 2 1

ISBN 0-387-98749-5 Springer-Verlag New York Berlin Heidelberg   SPIN 10710178

Foreword

For many years I have given talks to undergraduates on selected items in mathematics which could be extracted at a level understandable by students who have had calculus. I gave such talks regularly in Hirzebruch's classes in Bonn, at the Université de Montreal invited by Schlomiuk, a couple of times at the École Normale Supérieure in Paris a decade ago, more recently at the Humboldt University in Berlin invited by Kramer; and regularly during the 1980s invited by Wustholz at the ETH in Zurich. I am very grateful to U. Stammbach who translated the Zurich lectures in German for publication in *Elemente der Mathematik* from 1992 to 1997. I gave such students talks last to the Mathematics Undergraduate Student Association at Berkeley in the summer of 1998, and at the University of Texas at Austin in November 1998.

I am now presenting a collection of talks for undergraduates as a book. Such talks could be given by faculty, but even better, they may be given by students in seminars run by the students themselves. Such talks have been given at Yale, for instance.

In the United States, there is a need for remedial mathematics for a substantial group of students, and there is also a need to improve the school teachers' knowledge of mathematics, as well as their understanding of effective ways to support math learning at all levels. Some distinctions need to be made between the elementary schools and high schools. The situation involves many factors. Someone directly involved in elementary school teaching has emphasized to me the damage to mathematical learning and education already done by the time students reach high schools. I am speaking here in general terms of course, individual cases have to be evaluated on their own merits.

Under some legitimate social forces, and the above needs, certain excesses have taken place, whose effect is to lower standards and to impose a jargon as a precondition to get funding for various educational projects. More than a decade ago, the phrase was coined that mathematics should be "a pump, not a filter". Well, a few years ago in a program intended to stimulate mathematically gifted high school students, David Rohrlich extracted some material on p-adic L-functions which could be understood by high school seniors. My question at the time was: Are p-adic L-functions a pump or a filter? *Mutatis mutandis*, one can ask the same question about

the topics chosen for the present book. Is the Riemann hypothesis a pump
or a filter? Is the semi parallelogram law a pump or a filter? Is the study
of harmonic functions a pump or a filter?

The expression "A Pump, Not a Filter" is a section title p. 6 of the
NAS publication *Everybody Counts*. It was originally coined by Robert
M. White, at the time President of the National Academy of Engineering,
who said: "Mathematics must become a pump rather than a filter in the
pipeline of American education." I was once asked by an educational outfit
in Massachusetts to join a mathematical education project. A write up for
the project was headed "Connected Geometry—Priming the Pump", and
was written in the corresponding jargon. I asked those guys if they had to
write that way to get funded, and the answer on the phone was "yes". I
refused to join. This doesn't mean I am not interested in math education,
at all levels. Cf. my *Math! Encounters with High School Students* and *Geometry: A High School Course* (with Gene Murrow). As for the *Geometry*
book, I once got a letter from a teacher (also Department Head) at a girls
school. She wrote me:

> Enclosed are letters some of my ninth grade honors geometry class
> wrote to you. They are curious to meet the author of their textbook,
> and besides introducing themselves they were inviting you to come
> visit the class if you are ever in the . . . area.
>
> This is the first year I have taught from your geometry text, and
> as some of the girls are confirming, they find it quite challenging. I
> find it refreshing. You have converted me to teach 'paragraph' style
> proofs. Your constructions lend themselves well to Geometryssketch-
> pad, which we use frequently. The language of the book is rigorous,
> but clear and to the point. The textbook is perfect for our honors level
> geometry classes."

Her letter was accompanied by letters from girls in her class.

Some politicians, and as a result the National Science Foundation, have
expressed explicitly the point of view that there is enough money to fund
only research which has immediate applications, and not "curiosity driven"
research. I object. The rhetoric mixes up several issues which deserve being
disentangled. Among immediate applications is the need to have a competent mathematics teaching corps at all levels. Of course, many teachers are
competent, but there is also widespread incompetence, in addition to having some systematically disadvantaged districts. Then there are possible
uses of mathematics in engineering, biology, ecology, economics, whatever, which is quite a different thing. Where the money goes is legitimately
a political decision. But the slur against "curiosity driven" research as if
such research is useless (on whatever count) is unwarranted. Throughout
history, curiosity has led to remarkable discoveries of all types, and to some
extent most societies (the United States being a frequent exception) have
placed cultural value on scientific curiosity, ranging from China and India

to North Africa and Europe. Most of the discoveries actually turned out to have so-called applications, sometimes soon, sometimes only decades or centuries later. Anyhow, thinking about mathematics is very pleasurable to many people, and there is considerable evidence that we are programmed naturally to like mathematics, until the pleasure is ruined by incompetent teaching or other social factors. Every 5 year old kid I have ever met likes to add numbers, or subtract numbers. What happens afterwards is something else. The something else also may include the case of the teacher and the girls in her geometry class mentioned above, with their love of geometry and their enthusiasm. To repeat, individual cases have to be evaluated on their own merits. How many kids are lucky enough to make it to such a class?

Officials, in the government and in the universities, impose a jargon, involving pumps and filters or similar expressions confirming that one is subservient to the current social and political orthodoxy. The latest such jargon is the terminology of "vertically integrated" research and teaching and the "VIGRE" grants, "VI" meaning "vertically integrated". Are the talks in this book vertically integrated? Or horizontally? Or at what angle? Who's kidding whom? I wish it was a matter of kidding; actually it's a matter of funding.

So, never mind pumps, filters, and vertical integration. I hope under-graduates, or even some high school students, will enjoy these talks, or maybe some of them. Different students with different backgrounds need different things at different times for different purposes. Even among the talks I offer here, students with different backgrounds in mathematics will find some talks easy and others more difficult or impossible. So just pick and choose. It beats getting stoned, and it'll keep you off the streets.

Serge

New Haven 1998

Contents

Prime Numbers

I don't know any subject other than number theory where one can give a talk on mathematics centered around major unsolved problems, but understandable with almost no background in mathematics. High school students with slight calculus background should be able to understand most of this talk on prime numbers. The presentation is reworked from talks I have given to several audiences.

Here and some other places, I have kept some of the style of original talks, when students are questioned or interrupt freely. My book *MATH! Encounters with High School Students* (Springer-Verlag) was taken off tapes, literally, and is therefore entirely in the spoken style. Halmos once characterized this style as "vulgar", and obstructed publication of excerpts in the *Math Monthly*. A decade later, in the 1990s, the present talk was offered for publication again in the *Math Monthly*, and was turned down by the editor (Roger Horn, this time) because of the spoken style. Well, I like the spoken style, and I find it effective. Go figure.

Prime Numbers

Who knows what a prime number is?

[*Various students from the audience give a definition.*]

OK, I'll write down the definition. A prime number p is an integer ≥ 2 such that a divisor of p is either 1 or p itself. By convention, we take $p \geq 2$. It is agreed that 1 is not a prime number. So what is the sequence of prime numbers? It is

$$2, \ 3, \ 5, \ 7, \ 11, \ 13, \ 17, \ 19, \ 23, \ldots, \text{etc.}$$

What does "etc." mean? Does the sequence of prime numbers go on indefinitely, or will it stop? In other words: are there infinitely many primes or only a finite number of primes? The answer has been known since Euclid:

Theorem. *There are infinitely many prime numbers.*

Do you know how to prove it?

[Some students say they don't know a proof. One student gives a proof.]
Yes, that's a proof, and I shall write it down. Let 2, 3, . . . , P be prime numbers up to some prime P. What we have to do is to show that there is another prime number which is different from 2, 3, . . . , P. If we can show this, then we have shown that there are infinitely many prime numbers. So we let

$$N = (2 \cdot 3 \cdot 5 \cdots P) + 1.$$

If N is prime, then N is bigger than P, and we have found another prime. If N is not prime, then what happens? In that case, N is divisible by some prime. In fact, N factors as a product of two smaller positive integers. If one of them is a prime, we have already found a prime q dividing N. If not, we can factor these integers again into a product. Thus you can continue factoring, and you cannot continue indefinitely because the factors get smaller. Hence at some point you reach a prime factor q of N. Now the assertion is, that a prime factor q of N must be different from 2, 3, . . . , P. Why is that true? (*Some students in the audience give the answer.*) Yes, if you divide N by any of the primes 2, 3, . . . , P then you have a remainder 1, but q is a prime which divides N without a remainder, and so q cannot be equal to any of the primes 2, 3, . . . , P. So we have proved that there are infinitely many primes.
The proof dates back to Euclid.

Counting

We want to count how many primes there are. Since there are infinitely many, we have to make more precise what we mean. Given a positive number x, we let

$$\pi(x) = \text{number of primes which are} \leq x.$$

We want to estimate this number $\pi(x)$. Such an estimate depends on a probability. The general philosophy first conjectured by Gauss is that the probability that n is a prime is $1/\log n$. Gauss wrote this in a letter to the astronomer Encke in 1849, and also wrote that he had figured it out in 1792–1793, when he was 15 or 16 years old! Gauss' insight is the basic one in the theory of prime numbers. What does it mean? If you are given an integer n, it is either prime or not prime. We have to explain what we mean when we speak of the probability $1/\log n$. Gauss himself explained it. If you take the total number of primes less than or equal to x, $\pi(x)$, then $\pi(x)$ should simply be approximately the sum of the probabilities:

$$\pi(x) = \sum_{2 \leq n \leq x} \frac{1}{\log n} + E_1(x),$$

where $E_1(x)$ is some error term, which is small with respect to the value of the sum. The smaller the error term, the better is the approximation of $\pi(x)$ by the sum. To make everything precise, we should have an estimate for the error term, which makes it as small as possible compared to $\pi(x)$. We shall come to such an estimate in a moment, but first I want to point out that the sum can be rewritten another way, which will use calculus. Skip it if you don't know calculus.

Instead of writing the sum of $1/\log n$, you should have a tendency to write something else. Such a sum is approximately equal to what? Such a sum should remind you of something you have already seen: an integral. You should know this picture.

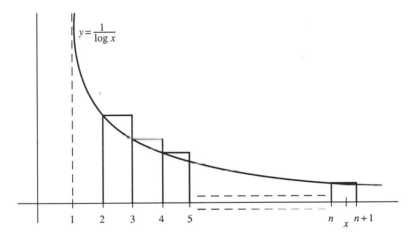

You should realize that the sum is a Riemann sum for the integral, so the sum is approximately like the integral of the function $1/\log x$, so we can write

$$\pi(x) = \int_2^x \frac{1}{\log t}\, dt + E_2(x),$$

where $E_2(x)$ is another error term; but $E_2(x)$ does not differ much from $E_1(x)$, in fact from calculus you can estimate that

$$\left| \int_2^x \frac{1}{\log t}\, dt - \sum_{2 \le n \le x} \frac{1}{\log n} \right| \le 2, \quad \text{for instance.}$$

So we can approximate $\pi(x)$ by the integral or the sum. In some ways the integral is nicer to handle because by integration by parts, we obtain

$$\int_2^x \frac{1}{\log t}\, dt = \frac{x}{\log x} + \int_2^x \frac{1}{(\log t)^2}\, dt - \frac{2}{\log 2}.$$

So on the right-hand side you are integrating something which is much
smaller than $1/\log t$, namely you are integrating

$$\frac{1}{\log t} \frac{1}{\log t},$$

which tends to 0 faster than $1/\log t$ as $t \to \infty$. This means that our expression for $\pi(x)$ can be written

$$\pi(x) = \frac{x}{\log x} + E_3(x),$$

with an error term $E_3(x)$ such that

$$\lim_{x \to \infty} \frac{E_3(x)}{x/\log x} = 0.$$

So $x/\log x$ gives a first order approximation to $\pi(x)$.

But $x/\log x$ is a not-so-good approximation of $\pi(x)$, because instead of
taking the good sum you have simply taken the total number of integers
from 2 to x, which is essentially x, and multiplied by the highest probability, which is $1/\log x$. You have done something which is not so good.
Taking the sum of the probabilities, or the integral, should be better, because in dividing x by $\log x$, you have averaged out too much. But the
expression $x/\log x$ has one advantage: it's a nice formula, in closed form,
with simple functions x and $\log x$. The formula gives a rough answer to
our first question. This answer was proved at the end of the last century.

Prime number theorem (Hadamard, de la Vallée-Poussin, 1896). *We
have*

$$\pi(x) \sim x/\log x \qquad for \quad x \to \infty.$$

The twiddle $\sim$ here has a precise meaning. If f and g are two functions of
a real variable x, we write

$$f(x) \sim g(x) \qquad for \quad x \to \infty$$

to mean that

$$\lim_{x \to \infty} f(x)/g(x) = 1.$$

This being the case, we say that $f(x)$ **is asymptotic to** $g(x)$ for $x \to \infty$.
Thus $\pi(x)$ is asymptotic to $x/\log x$.

The prime number theorem gives a first approximation of how many primes there are. The proof is—even today—not so easy. If you have had a course in complex analysis (residues, contour integration), then you can understand the proof which may take two hours—which is not so bad. The big theorems of complex analysis (Riemann mapping theorem, for instance) take about the same amount of time to prove. Today, the shortest proof is essentially Newman's proof [New 80]. (Cf. the last chapter of [La 93].)

The Riemann hypothesis

We come to the basic question: How good is the error term? Probabilistically, suppose you have random distribution. When you take the sum approximating $\pi(x)$,

$$S(x) = \sum_{2 \leq n \leq x} \frac{1}{\log n} \qquad \text{so} \qquad \pi(x) = S(x) + E_1(x),$$

or the integral

$$\pi(x) = \int_2^x \frac{1}{\log t}\, dt + E(x) = \mathrm{Li}(x) + E(x),$$

where $\mathrm{Li}(x)$ denotes the integral,

$$\mathrm{Li}(x) = \int_2^x \frac{1}{\log t}\, dt$$

and $E(x)$ is our error term, what should be an estimate for the error term $|E(x)|$? The conjecture in the probabilistic model is that there exists a constant C such that

$$|E(x)| \leq C(x^{1/2} \log x) \quad \text{for all } x \text{ sufficiently large.}$$

Since the main term is $x/\log x$ and the error term is bounded in absolute value by $C(x^{1/2} \log x)$, the error is rather small compared with $x/\log x$. It is quite easy to see that if the constant C exists, then it is small.

I am indebted to Hugh Montgomery for pointing out that there is some literature in number theory on that constant. He referred especially to [Scho 76], where it is proved that for $x > 2657$, one has, assuming the existence of the constant C above,

$$|E(x)| = |\pi(x) - \mathrm{Li}(x)| \leq \frac{1}{8\pi} x^{1/2} \log x.$$

The values for $x \leq 2657$ can be easily tabulated, and you should do so to get a feeling how good is the approximation of $\pi(x)$ by $\mathrm{Li}(x)$. The estimate $|E(x)| \leq C(x^{1/2} \log x)$ is the **Riemann Hypothesis**, which by consensus is the greatest conjecture in mathematics. Thus the problem of estimating the number of primes is to give a precise meaning to the probabilistic model, by describing the error term as accurately as possible.

In probabilistic terms, one could say that the Riemann Hypothesis states that the sequence of primes behaves like a random sequence in the probabilistic model that the probability for n to be prime is $1/\log n$.

There are other phenomena in number theory (and elsewhere) which can be phrased in a similar way. One makes up a probabilistic model, and by the law of large numbers, one knows the error term for a random sequence. Then a possible conjecture is that in a given specific case, a given sequence such as the sequence of primes behaves like a random sequence in that model. [See [LaT 76] for such probabilistic models in other contexts of number theory.]

We have stated the Riemann Hypothesis without knowing anything except the definition of a prime number and $\log x$.

COMMENT FROM THE AUDIENCE. That's not the way the Riemann Hypothesis is usually stated.

SERGE LANG. You're right, but the way it is usually stated takes more background to understand and is more complicated, because it uses more complex analysis. Since you mentioned it, I'll make a few comments on that. Euler defined the zeta function

$$\zeta(s) = \prod_p \left(1 - \frac{1}{p^s}\right)^{-1}$$

with the product taken over all prime numbers, and s real > 1. Riemann showed that there is an analytic function of a complex variable which has the Euler value for s real > 1, and is uniquely determined by these values. The Riemann Hypothesis originally stated that all the zeros of $\zeta(s)$ with $0 < \mathrm{Re}(s) < 1$ lie on the line $\frac{1}{2} + it$, i.e. on the line $\mathrm{Re}(s) = \frac{1}{2}$. If you formulate the Riemann Hypothesis that way, it may take about two hours (after a course in complex analysis) to show that this formulation is equivalent with the one we have given in terms of an estimate for the error term $E(x)$.

Mentioning complex analysis gives us an opportunity to comment on the way mathematics progresses. Hadamard developed certain parts of complex analysis motivated precisely by the desire to prove the prime number theorem, that is the asymptotic relation $\pi(x) \sim x/\log x$. Thus a problem in pure number theory gave rise to other theories which are used in most of the rest of mathematics, and physics, and engineering, and wherever else.

Conditions on primes

Now we leave the question about estimating the error term, and we discuss the counting problem when we put additional conditions on the primes. Let's start with specific examples. In the sequence of primes, you have pairs such as (3, 5), (5, 7), (11, 13), (17, 19), for which:

$$5 = 3 + 2;$$
$$7 = 5 + 2;$$
$$13 = 11 + 2;$$
$$19 = 17 + 2.$$

If you have a pair $(p, p + 2)$ such that p and $p + 2$ are prime, then you call these primes **twin primes**. What is the next one after (17, 19)? (Some students speak out the answer:)

$$(29, 31); \ (41, 43); \ (59, 61); \ (71, 73); \ldots$$

So what is the next question which we raise?

QUESTION. Are there infinitely many twin primes?
Who says "yes"? (*Some hands are raised.*)
Who says "no"? (*A few hands are raised.*)
(*Some students start to try to give a proof.*) I did not ask you how to prove it, I asked you whether you think there are infinitely many twin primes or not. That is a different question. At first, with the primes themselves, I asked you if you thought there are infinitely many. Then everybody thought yes, but you did not know how to prove it. So there are two questions. One is: "Do you think there are infinitely many?" and the other one is: "Can you prove it?" Right now we are at the first question. Do you think there are infinitely many twin primes or not? Does anybody think there are not infinitely many, that is there is only a finite number?

[*The audience keeps silent.*]

So it's too dangerous to speak out? Or are you thinking? All right, I'll give you the answer. The conjecture is "yes", but the proof is not known! This is the story at the moment: it is not known whether there are infinitely many twin primes. It is a famous problem of mathematics to answer this question. If you answer it, you'll make it in the history books.

Let's look at another question. You look for instance at $n^2 + 1$, and ask if such a number is prime. For instance:

$$2^2 + 1 = 5 \text{ is prime;}$$
$$4^2 + 1 = 17 \text{ is prime;}$$
$$6^2 + 1 = 37 \text{ is prime;}$$
$$8^2 + 1 = 65 \text{ is not prime;}$$
$$10^2 + 1 = 101 \text{ is prime.}$$

With the exception of 65, these numbers are prime. So what is the next question I will ask? (*Some students answer.*) Yes, the next question is: Are there infinitely many primes of type $n^2 + 1$?

How many think there are? (*Some hands are raised.*)
How many think there are not? (*A few hands are raised.*)
How many keep a prudent silence? (*Most keep silent.*)

I'll give you the answer: the conjecture is also "yes", but the proof is not known. This is another unsolved problem.

What arguments can one give to make it plausible that there are infinitely many? I will give you arguments which not only allow you to guess that there are infinitely many, but to guess how many. So we raise the following questions:

How many twin primes p, $p + 2$ are there with $p \leq x$?

How many primes p of type $n^2 + 1$ are there with $p \leq x$?

Take the question about twin primes. Suppose you grant that the probability for the number n to be prime is $1/\log n$. Can you guess what is the probability that n, $n + 2$ are twin primes?
[*A student does speak out the correct answer, after some fumbling.*]
Yes, the probability should be essentially the product

$$\frac{1}{\log n} \frac{1}{\log(n + 2)}.$$

But since $\log(n + 2)$ is very close to $\log n$ for n large, we can write $1/(\log n)^2$ instead, and so if we denote by $\pi_{tw}(x)$ the number of twin primes among the integers $2, 3, 4, 5, \ldots, x$, then conjecturally we should have

$$\pi_{tw}(x) \text{ is approximately } \frac{x}{(\log x)^2}.$$

But when I write the expression on the right, I am assuming implicitly that the probabilities are independent! Are they independent? The answer is NO. So we must be careful how we write the formula. The answer is that the number of twin primes $\pi_{tw}(x)$ satisfies the asymptotic formula

$$\pi_{tw}(x) \sim C_{tw}\frac{x}{(\log x)^2}, \qquad \text{or better} \qquad \sim C_{tw} \int_2^x \frac{1}{(\log t)^2} \, dt.$$

where C_{tw} is a positive constant, the twin prime constant. This constant reflects probabilistic dependencies. It is then a problem to compute this constant. We want an explicit formula for the constant, which is somewhat complicated, and is due to Hardy–Littlewood [HaL 23].

[*In order not to interrupt the main flow of ideas with a more technical question, a discussion of such constants is postponed till later.*]

Next we come to the question: What is the asymptotic estimate for the number $\pi_Q(x)$ of primes $p \leq x$ such that p can be written in the form $k^2 + 1$ with some positive integer k? Using the same heuristic technique, we have to describe the probability that n is of the form $k^2 + 1$, and is a prime number. This probability should roughly be

$$\frac{1}{\log n} \cdot \frac{1}{\sqrt{n}}.$$

Note that $\log \sqrt{x} = \frac{1}{2} \log x$, and so up to a constant factor, it does not matter whether we write $\log n$ or $\log \sqrt{n}$. We may then form the sum

$$\sum_{2 \leq n \leq x} \frac{1}{\log n} \frac{1}{\sqrt{n}} \quad \text{or the integral} \quad \int_2^x \frac{1}{(\log t) t^{1/2}} \, dt.$$

Integrating by parts, and guessing that there should be some constant factor C_Q to take into account more hidden probabilistic dependencies, we find conjecturally that

$$\pi_Q(x) \sim C_Q \frac{\sqrt{x}}{\log x}.$$

Hardy–Littlewood also determined the value of the constant C_Q (conjecturally).

Big generalization: The Bateman–Horn conjecture

Now we go on with a big generalization which systematically predicts the asymptotic behavior for prime counting in a very general case, including the special cases mentioned above. Let $f(T)$ be a polynomial with integer coefficients. We ask whether the values $f(n)$ are prime for infinitely many positive integers n. Write the polynomial in terms of its coefficients

$$f(T) = a_d T^d + a_{d-1} T^{d-1} + \cdots + a_0,$$

so $a_0, \ldots, a_d$ are integers, and a_d is called the leading coefficient. The integer d is called the degree if $a_d \neq 0$. We assume this is the case, and also $d \geq 1$, so the polynomial is not constant. What are necessary conditions that f represents infinitely many primes? Obviously, we must assume:

1. $f(T)$ is irreducible.
2. The leading coefficient a_d is positive.
3. The coefficients of $f(T)$ do not have a common prime factor.

Is that enough?

[Some students say yes, some keep a prudent silence. Once at UC Berkeley, two professors in the audience said yes. Then I pointed to a student, and asked: "What do you say?" The student said "sure".]

Well, the answer is NO! It was already realized by Bouniakowski in 1854 that it is not enough. For instance, the polynomial

$$f(T) = T^3 - T - 3$$

has values divisible by 3, i.e. $f(n)$ is divisible by 3 for all integers n. This is an elementary theorem of number theory, Fermat's (little) theorem, which says even more generally that for every prime number p, we have

$$n^p \equiv n \mod p \quad \text{for all integers } n.$$

In other words, $n^p - n$ is divisible by p for all integers n, but still the coefficients of $T^3 - T - 3$ or $T^p - T - p$ are relatively prime [*Pointing to the student who said "sure"*:] So it's not so sure, is it? Don't do what you did, saying "sure" based on what someone else says. Use your own brains. If you don't know, or want to think about it, say so. You won't do it again?

THE STUDENT. No, I won't.

SERGE LANG. OK. We go on. Bouniakowski saw how to fix things up. In 1854, he made the following conjecture [Bou 1854]. Let f be a polynomial with integer coefficients, of degree ≥ 1, and satisfying the following three conditions:

– The leading coefficient is positive.
– The polynomial is irreducible.
– No prime number divides all the values $f(n)$ when n ranges over the positive integers.

Then there exist infinitely many integers n such that $f(n)$ is prime.

This conjecture was generalized to several polynomials by Schinzel [SchS 58]. Even more than that, Bateman and Horn have given a quantitative form to the conjecture, which we now state precisely.

Let $f_1, \ldots, f_r$ be polynomials in one variable with integral coefficients and positive leading coefficient. Let $d_1, \ldots, d_r$ be their degrees. Let f be the product, $f = f_1 \cdots f_r$. Suppose:

– each polynomial is irreducible over the rational numbers;

– the polynomials are pairwise relatively prime, i.e. $(f_i, f_j) = 1$ for $i \neq j$;
– no prime number divides all the values $f(n)$ when n ranges over the positive integers.

We write vector notation, so $(f) = (f_1, \ldots, f_r)$. Let

$\pi_{(f)}(x) =$ number of positive integers $n \leq x$ such that
$f_1(n), \ldots, f_r(n)$ are all primes.

(We ignore the finite number of values of n for which some $f_i(n)$ is negative.) Observe that Bateman–Horn give an alternate way of counting. In the previous examples we counted the number of primes $p \leq x$ satisfying various conditions. Here we count the number of positive integers $n \leq x$ such that $f_i(n)$ is prime for each $i = 1, \ldots, r$. Bateman–Horn conjecture an asymptotic relation for $\pi_{(f)}(x)$ [BaH 62], namely

$$\pi_{(f)}(x) \sim (d_1 \cdots d_r)^{-1} C(f) \int_2^x \frac{1}{(\log t)^r} dt,$$

where $C(f)$ is an infinite product taken over all primes, that is

$$C(f) = \prod_p \left\{ \left(1 - \frac{1}{p}\right)^{-r} \left(1 - \frac{N_f(p)}{p}\right) \right\},$$

and $N_f(p)$ is the number of integers n with $1 \leq n \leq p$ such that $f(n)$ is divisible by p. If you know the language and notation of congruences, then we can also say that $N_f(p)$ is the number of solutions mod p of the congruence

$$f(n) \equiv 0 \mod p.$$

[If $N_f(p) = p$ for some prime p, then $C(f) = 0$ and

$$\pi_{(f)}(x) \leq d_1 + \cdots + d_r \quad \text{for all } x;$$

we agree to exclude this trivial case.] Bateman and Horn give heuristic justification for their conjecture, including machine computations.

I regard it as a major problem to give an estimate for the error term in the Bateman–Horn conjecture similar to the Riemann hypothesis. This could possibly lead to a vast reconsideration of the context for Riemann's explicit formulas.

All the examples we gave previously are special cases of the Bateman–Horn conjecture.

Suppose $r = 1$, and $f(T) = T$. Then the conjecture is just the prime number theorem.

Let $r = 1$ and $f(T) = T^2 + 1$. Then the conjecture actually gives the asymptotics discussed for the number of primes of the form $n^2 + 1$. Because Bateman–Horn count $n \leq x$ rather than $p \leq x$, it takes a few lines to show that their asymptotic answer is the same as the answer found by Hardy–Littlewood. I was curious once about the relation between the two variations, and I carried out the computation, which took me about a page.

Let $r = 2$ and let $f_1(T) = T$, $f_2(T) = T + 2$. Then the twin prime counting conjecture comes out of the Bateman–Horn conjecture. It also takes about a page in this case to show that the Hardy–Littlewood constant is the same as the Bateman–Horn constant.

Finally, going back to one polynomial, let a, b be integers, $a > 0$, and a, b relatively prime. Let $f(T) = aT + b$. Then the Bateman–Horn conjecture gives the fact that there are infinitely many primes in an arithmetic progression, together with the density of such primes, which is actually a famous theorem of Dirichlet.

The expression for the constant $C(f)$ in front of the logarithmic integral constitutes an example of a very general phenomenon occurring frequently in the theory of diophantine equations. The density of the solutions is expressed by a product of p-densities for each prime p, and another real factor which in the Bateman–Horn conjecture is $1/(d_1 \cdots d_r)$. The factor which contains the prime p describes the behavior modulo p, i.e. the behavior with respect to divisibility by p, and in particular, this factor contains the number of solutions of the congruence modulo p. In [HaL 23] and [LaT 76] you will find further examples of this phenomenon.

Appendix

Comments on the constants

The Bateman–Horn conjecture gave a systematic answer for the constant appearing in front of the logarithmic integral. In several special cases, Hardy–Littlewood had computed the constant [HaL 23], but using the alternate way of counting, namely primes $p \leq x$ rather than $n \leq x$. We now return to a discussion of the Hardy–Littlewood constants. They are expressed as infinite products.

We start with the twin prime constant C_{tw}. The story of that constant is quite interesting. Sylvester in 1871 and Brun in 1915 guessed the wrong constant because they did not take into account some hidden probabilistic dependences between the primes. Hardy and Littlewood in 1923 [Hal 23] remarked that Sylvester and Brun had conjectured a false formula, and themselves conjectured the correct value of the constant C_{tw}. "Correct" means partly that you can make a table giving the values for $\pi_{tw}(x)$ and the values for $C_{tw}x/(\log x)^2$, and compare, and that this empirical verification justifies the stated value. So "correct" is used here partly in the sense of physics. But there is another component to being "correct" (lacking an actual proof), namely that the conjecture fits certain strong structural heuristics, that there is a definite coherent structure from which the guess follows, the structure being itself partly conjectural. Hardy and Littlewood also had this structure. Of course, it may not be the only structure. Anyhow, the value given by Hardy–Littlewood is

$$C_{tw} = 2 \prod_{p \text{ odd}} \left(1 - \frac{1}{(p-1)^2}\right).$$

The product is taken over all primes $\neq 2$, that is over all odd primes. As you can see, this constant looks somewhat different from the Bateman–Horn constant. The difference is due to the alternate ways of counting. Since Hardy–Littlewood and Bateman–Horn have very different approaches to the constant, but they find the same answer, one regards this as confirmation that the Hardy–Littlewood answer is correct, because it fits structurally in other investigations.

Next we give the Hardy–Littlewood value for the constant C_Q coming into the conjectured asymptotic formula for the number of primes $\leq x$ of the form $p = n^2 + 1$. The conjectured value is

$$C_Q = \prod_{p \text{ odd}} \left(1 - \chi(p)\frac{1}{p-1}\right),$$

where $\chi(p) = 1$ if -1 is a square mod p, i.e. if $n^2 + 1 \equiv 0 \bmod p$ has a solution; and $\chi(p) = -1$ if -1 is not a square mod p. [See also [LaT 76], Part II, especially p. 81, Example 1, where this constant is derived from an entirely different probabilistic model.]

[*A student asks how one arrives at such conjectured formulas, and how one makes guesses.*]

Yes, let me deal with that question. First, there is an unexplainable component, namely the brain of a mathematician who knows how to look and where to look and is good at guesses. Having such talent is one of the distinguishing factors between a research mathematician and other people. Then there is the hard work which follows, how one checks the guesses and how one structures them, to write a research paper.

Let's deal with the question how one may arrive at the idea that there are infinitely many twin primes, or primes of the form $k^2 + 1$. Do you regard what we have done as some sort of an answer? Yes? So-so? Well, it's something, it gives you not only that there are infinitely many, but it gives you a qualitative measure for how many. One could push the probabilistic model to make it finer, so that you also get the value of the constant, to see more precisely how many there should be.

It's actually not so easy to go all the way, as you can see from the fact that first rate mathematicians like Sylvester and Brun didn't quite get it right the first time around. So the Hardy–Littlewood paper is a very good paper.

I will show you now one of the approaches that eventually would lead to the constants. Things of necessity have to get somewhat more complicated. We'll deal just with the ordinary case of the prime number theorem.

You take all the numbers $1, 2, 3, 4, 5, 6, \ldots, x$. How many are divisible by 2? Approximately $x/2$. The number of integers $\leq x$, *not* divisible by 2, is approximately $(1 - \frac{1}{2})x$, and the fraction of x is $(1 - \frac{1}{2})$. The number of integers *not* divisible by 3 is essentially

$$(1 - \tfrac{1}{3})x.$$

The number of integers *not* divisible by 5 is

$$(1 - \tfrac{1}{5})x.$$

So, to get the number $\pi(x)$ of primes $\leq x$, we want to know how many integers are not divisible by 2, not by 3, not by 5, etc. How many will you get? The answer is roughly:

$$G(x) \cdot x \qquad \text{where} \qquad G(x) = \prod_{p \leq x} \left(1 - \frac{1}{p}\right).$$

This is a beginning to determine the correct constant for $\pi(x)$, the number of primes $\leq x$, except that we have assumed that the conditions of non-divisibility are independent, and they are not. So we must take the fraction

$$\prod_{p \leq x} \left(1 - \frac{1}{p}\right) = G(x)$$

times some constant because of complicated conditions of dependence. The constant here is e^{γ}, where γ is a constant called the Euler constant. It is a theorem of Mertens that one has the asymptotic relation

$$e^{\gamma} G(x) \sim \frac{1}{\log x},$$

and therefore

$$\pi(x) \sim e^{\gamma} G(x) x \sim \frac{x}{\log x}.$$

You can find a discussion of why e^γ appears in the paper by Hardy–Littlewood, and also Hardy and Wright [HaW 60], Chapter 22, 22.20.

So you see, just to discuss the ordinary prime number theorem, it already becomes more technical and complicated. All the more for the precise probabilistic heuristics of Hardy–Littlewood for the twin primes, for the primes of the form $n^2 + 1$, and other cases. At this point, if you are still with it, go read both Bateman–Horn and Hardy–Littlewood.

Some graphs

I am indebted to Steve Miller for making up some graphs illustrating experimental behavior of the zeta function and of the three functions $x/\log x$, $\pi(x)$, and $\mathrm{Li}(x)$ in certain ranges.

The first graph is that of the function $\xi(\frac{1}{2} + it)$, which is necessarily real as a function of t because of the functional equation.

Graph of $\quad \xi(\frac{1}{2} + it), \qquad -50 < t < 50.$

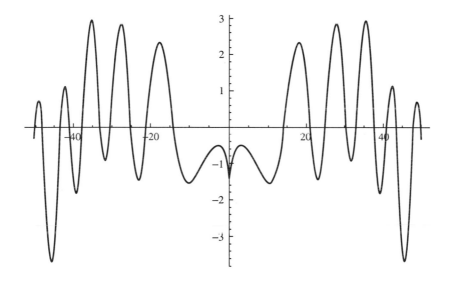

The next three graphs compare $x/\log x$, $\pi(x)$, and $\mathrm{Li}(x)$ over three different ranges for x, namely up to 100, 1000, and 10,000, respectively. Note that as one goes further, it becomes experimentally evident that $\mathrm{Li}(x)$ gives a better approximation to $\pi(x)$ than $x/\log x$. It is also "evident" that $\mathrm{Li}(x)$ lies above $\pi(x)$. To what extent can you conclude that this statistical behavior persists? Go ask some analytic number theorist. Or see Ingham [Ing 32], Chapter V, Theorem 35.

Graphs of Li(x), π(x), x/log x

Li(x) is the top graph.

$\pi(x)$ is the middle graph.

$\dfrac{x}{\log x}$ is the bottom graph.

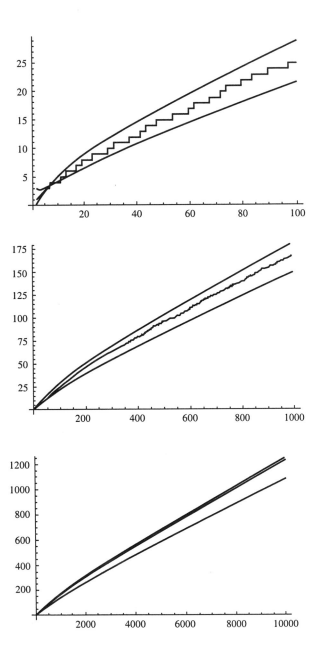

Bibliography

[BaH 62] P. BATEMAN and R. HORN, A heuristic asymptotic formula concerning the distribution of prime numbers, *Math. Comp.* **16** (1962) pp. 363–367

[Bou 1854] V. J. BOUNIAKOWSKY, Sur les diviseurs numériques invariables des fonctions rationnelles entières, *Mem. Sci. Math et Phys.* **T. VI** (1854) pp. 307–329

[Br 15] V. BRUN, Über das Goldbachsche Gesetz und die Anzahl der Primzahlpaare, *Archiv for Mathematik* (Christiania) **34** Part 2 (1915) pp. 1–15

[HaR 74] H. HALBERSTAM and H. RICHERT, *Sieve Methods*, Academic Press, 1974

[HaL 23] G. H. HARDY and J. E. LITTLEWOOD, Some problems of Partitio Numerorum, *Acta Math.* **44** (1923) pp. 1–70

[HaW 80] G. H. HARDY and E. M. WRIGHT, *An Introduction to the Theory of Numbers*, Fourth edition, Oxford University Press, 1980

[Ing 32] A. E. INGHAM, *The Distribution of Prime Numbers*, Cambridge University Press, 1932

[La 93] S. LANG, *Complex Analysis*, Third edition, Springer-Verlag, 1993 (Fourth edition, 1999)

[LaT 76] S. LANG and H. TROTTER, *Frobenius Distributions in* GL_2*-Extensions*, *Springer Lecture Notes* **504**, Springer-Verlag, 1976

[New 80] D. J. NEWMAN, Simple analytic proof of the prime number theorem, *Amer. Math. Monthly* **87** (1980) pp. 693–696

[Pat 88] S. J. PATTERSON, *An Introduction to the Theory of the Riemann Zeta Function*, Cambridge Studies in Advanced Mathematics, vol. 14, Cambridge University Press, 1988

[SchS 58] A. SCHINZEL and W. SIERPINSKY, Sur certaines hypothèses concernant les nombres premiers, *Acta Arith.* **4** (1958) pp. 185–208

[Scho 76] L. SCHOENFELD, Some sharper Tchebychev estimates II, *Math. Comp.* **30** (1976) pp. 337–360

[Syl 1871] J. SYLVESTER, On the partition of an even number into two prime numbers, *Nature* **56** (1896–1897) pp. 196–197 (= *Math. Papers* **4** pp. 734–737).

The *abc* Conjecture

Let's start with a theorem about polynomials. You probably think that one knows everything about polynomials. Most mathematicians would think that, including myself. It came as a surprise when R. C. Mason in 1983 discovered a new, very interesting fact about polynomials [Ma 83]. Even more remarkable, this fact actually had been already discovered by another mathematician, W. Stothers [Sto 81], but people had not paid attention and I learned of Stothers' paper only much later, from U. Zannier [Za 95] who also rediscovered some of Stothers' results. So the history of mathematics does not always flow smoothly.

Anyhow, although history isn't bunk, let us leave history aside to state and prove the Mason–Stothers theorem. After that, we'll discuss how other mathematicians have translated this theorem into a conjecture about ordinary integers. We consider polynomials with complex numbers as coefficients. The set of all such polynomials in a variable t is denoted by $\mathbf{C}[t]$. We write a non-zero element of $\mathbf{C}[t]$ in the form

$$f(t) = c_1 \prod_{i=1}^{r} (t - \alpha_i)^{m_i},$$

where $\alpha_1, \alpha_2, \ldots, \alpha_r$ are the distinct roots of f, and c_1 is a constant, $c_1 \neq 0$. The integers m_i $(i = 1, \ldots, r)$ are the multiplicities of the roots, and the degree of the polynomial f is

$$\deg f = m_1 + \cdots + m_r.$$

The number of (distinct) roots of f will be denoted by $n_0(f)$, so by definition,

$$n_0(f) = r.$$

It's obvious that $\deg f$ can be very large, but $n_0(f)$ may be small. For instance, $f(t) = (t - \alpha)^{1000}$ has degree 1000, but $n_0(f) = 1$. If f, g are two non-zero polynomials, then in general

$$n_0(fg) \leq n_0(f) + n_0(g).$$

If in addition f, g are relatively prime, then we actually have equality

$$n_0(fg) = n_0(f) + n_0(g).$$

The Mason–Stothers theorem states:

Theorem. *Let* $f, g, h \in \mathbf{C}[t]$ *be non-constant relatively prime polynomials satisfying* $f + g = h$. *Then*

$$\max(\deg f, \deg g, \deg h) \leq n_0(fgh) - 1.$$

The theorem shows in a very precise way how the relation $f + g = h$ implies a bound for the degrees of f, g, h, namely the number of distinct roots of the polynomial fgh, even with -1 tacked on.

Before we give Mason's proof, we mention some applications which show how powerful the theorem is. You all know Fermat's last theorem [which was proved by Wiles [Wi 95] a decade later]:

Let n be an integer ≥ 3. There are no solutions of the equation

$$x^n + y^n = z^n$$

in non-zero integers x, y, z.

The analogous theorem for polynomials was known in the last century, and was proved by arguments from algebraic geometry. Here we shall give a much simpler proof by using the Mason–Stothers theorem.

Theorem. *Let n be an integer ≥ 3. There is no solution of the equation*

$$x(t)^n + y(t)^n = z(t)^n$$

with non-constant relatively prime polynomials $x, y, z \in \mathbf{C}[t]$.

Proof. Let $f(t) = x(t)^n$, $g(t) = y(t)^n$, and $h(t) = z(t)^n$. Then the Mason–Stothers theorem yields

$$\deg x(t)^n \leq n_0(x(t)^n y(t)^n z(t)^n) - 1.$$

However $\deg x(t)^n = n \cdot \deg x(t)$ and $n_0(x(t)^n) = n_0(x(t)) \leq \deg x(t)$. Hence

$$n \cdot \deg x(t) \leq \deg x(t) + \deg y(t) + \deg z(t) - 1.$$

Similarly, we obtain the analogous inequality for $y(t)$ and $z(t)$, that is:

$$n \cdot \deg y(t) \leq \deg x(t) + \deg y(t) + \deg z(t) - 1,$$
$$n \cdot \deg z(t) \leq \deg x(t) + \deg y(t) + \deg z(t) - 1.$$

Adding the three inequalities yields

$$n \cdot (\deg x(t) + \deg y(t) + \deg z(t)) \leq 3 \cdot (\deg x(t) + \deg y(t) + \deg z(t)) - 3.$$

Therefore

$$(n - 3)(\deg x(t) + \deg y(t) + \deg z(t)) \leq -3.$$

For $n \geq 3$ this is obviously impossible, because the left side is then ≥ 0 while the right side is negative. So we have proved Fermat's last theorem for polynomials.

The proof of Fermat's last theorem for polynomials is much more diffi-cult without using the Mason–Stothers theorem. It's not clear how long it would take you, or a random mathematician, to prove it, by whatever means. Try it out on your friends and see how long it takes them. The above proof with Mason–Stothers is simple and elegant. We shall give generalizations later. But having shown you the power of the theorem, we shall now prove it.

Proof of the Mason–Stothers theorem. In the statement of the theorem, we have the degree on the left side and the number n_0 on the right side, that is the number of distinct roots of a polynomial. So we have to find a way to understand and control the multiplicities of the roots. To do that, we divide the equation $f + g = h$ by h, and obtain

$$\frac{f}{h} + \frac{g}{h} = 1.$$

Put $R = f/h$, $S = g/h$. Then $R + S = 1$. Now you should have an irre-sistible impulse to do something to this equation. What do you do to func-tions? You take their derivatives. So we get the new equation $R' + S' = 0$, which we rewrite in the form

$$(1) \qquad \frac{R'}{R}R + \frac{S'}{S}S = 0.$$

Consider the quotient g/f. With our notation, and equation (1), we obtain

$$(2) \qquad \frac{g}{f} = \frac{S}{R} = \frac{R'/R}{S'/S}.$$

In this way, we have expressed g/f as a quotient of logarithmic derivatives. Indeed, in calculus, if F is a function, then F'/F is called its logarithmic derivative. By using this logarithmic derivative, we shall have the multiplicities of the roots under control. We need an immediate property of the logarithmic derivative, namely it transforms products into sums. In other words, for two functions F, G we have

$$\frac{(FG)'}{FG} = \frac{F'}{F} + \frac{G'}{G}.$$

This follows immediately for the rule for differentiation of a product. Then for quotients, we obtain

$$\frac{(F/G)'}{F/G} = \frac{F'}{F} - \frac{G'}{G}.$$

So the logarithmic derivative transforms quotients into differences. Then by induction, we obtain similar relations for sums or quotients of several factors. For example, if $f_1, \ldots, f_n$ are polynomials, then

$$\frac{(f_1 \cdots f_n)'}{f_1 \cdots f_n} = \frac{f_1'}{f_1} + \cdots + \frac{f_n'}{f_n}.$$

We apply this rule to the factorization of our polynomials f, g, h at the beginning, for which we have factorizations

$$f(t) = c_1 \prod (t - \alpha_i)^{m_i},$$
$$g(t) = c_2 \prod (t - \beta_j)^{n_j},$$
$$h(t) = c_3 \prod (t - \gamma_k)^{l_k}.$$

Note that the logarithmic derivative of a constant is 0. For a polynomial $(t - \alpha)$, the logarithmic derivative is $1/(t - \alpha)$. Thus taking the logarithmic derivative of $f(t)$, $g(t)$, $h(t)$ yields

$$\frac{f'}{f} = \sum \frac{m_i}{(t - \alpha_i)}, \qquad \frac{g'}{g} = \sum \frac{n_j}{(t - \beta_j)}, \qquad \frac{h'}{h} = \sum \frac{l_k}{(t - \gamma_k)}.$$

Applying the rule for the logarithmic derivative to $R = f/h$ and $S = g/h$, we get from (2):

$$(3) \qquad \frac{g}{f} = -\frac{f'/f - h'/h}{g'/g - h'/h} = \frac{\displaystyle\sum \frac{m_i}{t - \alpha_i} - \sum \frac{l_k}{t - \gamma_k}}{\displaystyle\sum \frac{n_j}{t - \beta_j} - \sum \frac{l_k}{t - \gamma_k}}.$$

We put things over a common denominator, so let

$$D(t) = \prod(t - \alpha_i) \prod(t - \beta_j) \prod(t - \gamma_k).$$

Obviously we have deg $D(t) = n_0(fgh)$. Hence

$$\deg\left(\frac{D(t)}{(t - \alpha_i)}\right) = n_0(fgh) - 1 = \deg\left(\frac{D(t)}{(t - \beta_j)}\right) = \deg\left(\frac{D(t)}{(t - \gamma_k)}\right).$$

We multiply the numerator and denominator of (3) with $D(t)$. We then obtain

$$\frac{g}{f} = -\frac{\displaystyle\sum \frac{m_i}{t - \alpha_i} - \sum \frac{l_k}{t - \gamma_k}\, D(t)}{\displaystyle\sum \frac{n_j}{t - \beta_j} - \sum \frac{l_k}{t - \gamma_k}\, D(t)}$$

$$= \frac{\text{polynomial of degree} \leq n_0(fgh) - 1}{\text{polynomial of degree} \leq n_0(fgh) - 1}.$$

Thus g/f is a quotient of two polynomials of degrees $\leq n_0(fgh) - 1$. Since f, g are relatively prime, it follows that also the degrees of g and f are at most $n_0(fgh) - 1$. Finally, since $h = f + g$, we conclude that deg h is at most $n_0(fgh) - 1$. This concludes the proof of the Mason–Stothers theorem.

It's remarkable that such a simple and yet powerful result on polynomials was discovered only in the early 1980s, late into the twentieth century.

Next, we are going to translate the theorem into a statement about integers. You may know already that there is a deep analogy between the integers and polynomials. For instance, both systems satisfy the euclidean algorithm, and so in both systems unique factorization into irreducible or prime elements holds. We look for a number associated to an integer, in such a way that it is an analogue of the number n_0 of a polynomial. We also look for a number analogous to the degree of a polynomial. Under multiplication, the degrees add, that is the degree of a product of polynomials is the sum of the degrees of the individual factors. Experience shows that the analogue of the degree is the logarithm of the absolute value of an integer. So for a prime number p, the analogue of the degree of an irreducible polynomial is just the logarithm $\log p$. For two positive integers m, n we have $\log(mn) = \log m + \log n$. If

$$m = \prod p_i^{m_i}$$

is the factorization of m into prime factors, then

$$\log m = \sum m_i \log p_i.$$

Next, we want to translate n_0. For a polynomial

$$f(t) = \prod (t - \alpha_i)^{m_i},$$

let us define

$$N_0(f) = \prod (t - \alpha_i),$$

so we replace the multiplicities m_i by 1 in the factorization of $f(t)$. Then by definition,

$$n_0(f) = \text{degree of } N_0(f(t)).$$

So what's the right definition of $n_0(m)$ for a positive integer m? Suppose

$$(4) \qquad\qquad m = p_1^{m_1} \cdots p_r^{m_r}$$

is the factorization of m into prime powers. What is $n_0(m)$?

A STUDENT. I think $n_0(m)$ should be the number of distinct prime factors, so $n_0(m) = r$.

SERGE LANG. This is not a bad answer, but it's not quite good enough. When we deal with complex polynomials, then they have unique factorization into primes, that is irreducible polynomials, but these have degree 1. If we dealt with polynomials with other types of coefficients, for instance polynomials over the rational numbers, then an irreducible polynomial does not necessarily have degree 1. Something analogous happens with prime numbers. This means that we have to give a weight to a prime number, and this weight is $\log p$. Hence for the positive integer m with the prime factorization (4), we define

$$n_0(m) = \sum_{i=1}^{r} \log p_i.$$

We can write this expression also in the form

$$n_0(m) = \sum_{p \mid m} \log p.$$

Then we define

$$N_0(m) = \prod_{p|m} p = \prod_{i=1}^{r} p_i.$$

Thus $N_0(m)$ is the product of the primes dividing m, but taken with multiplicity 1. One calls $N_0(m)$ the *radical* of m.

Now that we have a definition of $n_0(m)$ and $N_0(m)$, what is the analogue of the Mason–Stothers theorem? It's easy to translate the hypothesis:

Let a, b, c be non-zero relatively prime integers such that $a + b = c$.

Then we want an inequality of the form

$$\max(|a|, |b|, |c|) \leqq ???.$$

What should be on the right side of the inequality?

A STUDENT. Maybe $N_0(abc) - 1$?

SERGE LANG. It's not quite so simple. First about the -1: in the case of polynomials, it was a gift from the gods, and we really can't expect something so sharp for integers. Besides, we're working here multiplicatively, not additively. So let's look at the inequality

$$\max(|a|, |b|, |c|) \leqq N_0(abc).$$

Unfortunately, it turns out that this inequality in general is false. This was first noticed by Stewart–Tijdeman [StT 86], who also showed that it remains false if we replace the right side by a constant times $N_0(abc)$ no matter how large the constant is. In other words, *there is no constant K such that the inequality*

$$\max(|a|, |b|, |c|) \leqq K \cdot N_0(abc)$$

is true for all relatively prime integers a, b, c with $a + b = c$. Two Yale students, Wojtek Jastrzebowski and Dan Spielman gave me the following simple argument showing this.

We consider the equation $a_n + b_n = c_n$ with

$$a_n = 3^{2^n}, \qquad b_n = -1, \qquad \text{and} \qquad c_n = 3^{2^n} - 1.$$

Writing $3 = 1 + 2$, it's easy to prove by induction that 2^n divides $3^{2^n} - 1$.

Hence

$$N_0(a_n b_n c_n) \leqq 3 \cdot 2 \cdot \frac{c_n}{2^n}.$$

However, the inequality

$$3^{2^n} \leqq K \cdot 3 \cdot 2^{\frac{3^{2^n}}{2^n}}$$

cannot hold for all n no matter how large we select the constant K, because $k \cdot 3 \cdot 2/2^n$ approaches 0 as n gets larger and larger.

So we have to modify and weaken the inequality further. The right conjecture, called the **abc conjecture**, was made by Masser and Oesterle. It is one of the best conjectures of the century.

The *abc* conjecture (Masser, Oesterle, 1986). *Given $\epsilon > 0$, there exists a constant $K(\epsilon)$ such that for all non-zero relatively prime integers a, b, c with $a + b = c$, we have the inequality*

$$\max(|a|, |b|, |c|) \leqq K(\epsilon)(N_0(abc))^{1+\epsilon}.$$

Note that Stewart–Tijdeman give some lower bound for ϵ [StT 86].

Essentially with the same argument as for polynomials, one can show that the *abc* conjecture implies Fermat's last theorem for sufficiently large n. We give the argument in detail. Without loss of generality, we can assume that a, b, c are positive integers, so we don't have to write the absolute value signs. Let x, y, z be positive relative prime integers such that

$$x^n + y^n = z^n.$$

Let $a = x^n$, $b = y^n$, and $c = z^n$. Then

$$N_0(x^n y^n z^n) = N_0(xyz) \leqq xyz.$$

Applying the *abc* conjecture, we get

(5) $x^n \ll (xyz)^{1+\epsilon}, \qquad y^n \ll (xyz)^{1+\epsilon}, \qquad z^n \ll (xyz)^{1+\epsilon},$

where the sign $\ll$ abbreviates the inequality $\leqq K(\epsilon)$, with the constant $K(\epsilon)$ depending on ϵ. Thus the left side of the sign $\ll$ is $\leqq$ the product of $K(\epsilon)$ times the right side. Taking the product of the inequalities in (5), we obtain

$$(xyz)^n \leqq (xyz)^{3+3\epsilon}.$$

Taking the log, we get

$$(n - 3 - 3\epsilon) \, \log(xyz) \leqq \log K,$$

with the constant $K = K(\epsilon)$. Since $xyz \geqq 2$, this last inequality gives an upper bound for n, thus proving Fermat's last theorem for all sufficiently large n.

The bound for n depends on the constant $K(\epsilon)$. There is no conjecture today about giving an effective estimate for $K(\epsilon)$. Of course, just for effectivity, we can take $\epsilon = 1$, so an effective determination of $K(1)$ would give an effective proof for Fermat's last theorem, reducing it to computations for a finite number of cases. Naturally, Wiles' proof is valid for all n, but there are equations similar to Fermat's for which one doesn't know the analogue of Wiles' theorem, and an approach via the *abc* conjecture might work.

A STUDENT. Has anyone tried to make computations which might yield a counterexample to the *abc* conjecture?

SERGE LANG. That's not the way it goes. Tables of prime factorizations in various cases actually confirm the conjecture. Note that the exponent $1 + \epsilon$ in the exponent makes the conjecture very powerful. Among other things, the *abc* conjecture says that if there are primes with large exponents in the factorization of a, b, c, then these primes are compensated by many small primes occurring only with exponent 1. For example, there are tables for the prime factorization of $2^n \pm 1$ and similar numbers in the paper [BLSTW 83]. These tables show clearly that almost all primes occur only with exponent 1. If there are small primes with bigger exponents, then they are compensated by large primes with exponent 1.

A STUDENT. Does Fermat's last theorem imply the *abc* conjecture?

SERGE LANG. No. Fermat's last theorem is just a special case. The *abc* conjecture is much stronger and gives much more information about the way the exponents of the primes in a factorization of *abc* are bounded. To make the point even clearer, for Fermat's last theorem, we don't have to show the *abc* conjecture with the exponent $1 + \epsilon$. Any *fixed* exponent would suffice.

A STUDENT. How does one arrive at such a conjecture?

SERGE LANG. Masser and Oesterle didn't just hit upon the conjecture. They were working in a much larger context, which was not at all elementary. The way I presented the conjecture here does not reflect the way it developed historically. Real life is much more complicated. The conjecture arose from deep considerations of algebraic geometry and the theory of modular functions, not just in connection with the Mason–Stothers theorem. Those considerations are too complicated for me to describe them today. Still, I can make a few more comments which will open up some possibilities.

Consider an equation of the form

$$u^3 - v^2 = k$$

to be solved in relatively prime integers u, v, and k. This equation was first considered by M. Hall [Ha 71]. He made the following conjecture:

Hall conjecture. *For integers u, v, and k with $u^3 - v^2 = k \neq 0$, we must have*

$$|u|^3 \ll |k|^{6+\epsilon} \qquad and \qquad |v|^2 \ll |k|^{6+\epsilon}.$$

Actually, Hall made the conjecture without the ϵ! At the time, he didn't see the need for an ϵ. For a while, I thought an ϵ was needed. It's now not clear to me whether it is needed or not in the Hall conjecture. Maybe Hall was strictly correct in the first place because of the special kind of equation he was considering. In any case, "Hall's conjecture for polynomials" was already proved in 1965 by Davenport [Da 65], even in sharper form, without any undetermined constant in the estimate.

Davenport's theorem. *Let f, g be two non-constant polynomials with $f^3 - g^2 \neq 0$. Then*

$$\tfrac{1}{2} \deg f \leqq \deg(f^3 - g^2) - 1$$

and

$$\tfrac{1}{3} \deg g \leqq \deg(f^3 - g^2) - 1.$$

You can prove these inequalities directly by applying the Mason–Stothers theorem when the polynomials f, g are relative prime. It's a good exercise in algebra to prove the inequalities also in the case when f, g are not relatively prime. Then you have to get rid of common factors step by step, to reduce the theorem to the case when f, g are relatively prime. So you are led to consider a more general equation of the form

$$Af^3 + Bg^2 = h,$$

with polynomial coefficients A, B. The bound for the degrees of f, g, h in such an equation will then depend on A and B.

Let's go back to integers. We consider again the equation

$$u^3 - v^2 = k$$

with relatively prime u, v, k. Applying the *abc* conjecture, we get the bounds

(6) $\qquad |u|^3 \ll (N_0(k))^{6+\epsilon}$ and $\qquad |v|^2 \ll (N_0(k))^{6+\epsilon}$.

To carry out the details is an easy exercise. Hall's conjecture with epsilon then follows from the *abc* conjecture, at least for relatively prime u, v, k, because $N_0(k) \leq |k|$. More generally, consider the equation

$$Au^3 + Bv^2 = k,$$

with non-zero integer coefficients A, B. Then the *abc* conjecture gives bounds as in (6). Of course, the constants implicit in the inequalities (6) will then depend on A and B. Even more generally, we may consider equations of higher degree, specifically

$$Au^n + Bv^m = k.$$

The exponents n, m are positive integers. We assume that $mn \neq m + n$. Then it's an easy exercise to apply the *abc* conjecture to derive the inequality

$$|u|^n \ll (N_0(k))^{mn(1+\epsilon)/[mn-(m+n)]},$$

and similarly for $|v|^m$. As already mentioned, in this inequality the implicit constants depend on A and B. Going back to $m = 3$ and $n = 2$, there are special values of A and B which are especially interesting. For example $A = -4$ and $B = -27$. Then the expression

$$\Delta = -4u^3 - 27v^2$$

is well known to be the discriminant of the polynomial

$$X^3 + uX + v.$$

For these special values of A and B, the corresponding conjecture is the generalized Szpiro conjecture. Actually, Szpiro made the original conjecture not with the number N_0 but with a much more complicated invariant N, which arises in the theory of elliptic curves, that is the theory of equations of the form

$$Y^2 = X^3 + uX + v.$$

This theory is hard to explain at the level of the present discussion, and we don't go into it. In any case, Szpiro was led to his conjecture by deep considerations of algebraic geometry and number theory. Again, in so far as I used only the simple invariant N_0, which was easy to define, and in so far as we connected the Szpiro conjecture only with the Mason–Stothers theorem and the *abc* conjecture, I did not follow the historical order. In this sense I have been lying to you, because I swept away big mathematical fields which were important in the historical development of all the conjectures. From the whole set up, I extracted what could be explained in simple terms inside an hour. But the mathematicians who made these conjectures didn't come to them so directly and easily. Only after hard and extensive work with these deep and extensive theories. [*For more applications of the abc conjecture to the theory of elliptic curves, and further remarks, cf.* [La 90], *which contains a more extensive bibliography.*]

To conclude, I point out that the original Szpiro conjecture did not actually refer to inequality (6), but only to the weaker inequality

$$(7) \qquad\qquad |\Delta| \ll (N_0(\Delta))^{6+\epsilon}.$$

The stronger conjecture (6), which give bounds for $|u|$ and $|v|$, not only for $|\Delta|$, was formulated only later. That's why we referred to (6) as the generalized Szpiro conjecture.

So you see, it took a long time for the development of the *abc* conjecture to realize its central position in number theory and the theory of equations. The history did not follow a straight line, but went through detours and analogies, whereby Davenport's theorem and theorems about polynomials as well as Hall's conjecture played a role. But that's the way mathematics develops!

Appendix

[*In Spring 1998, I had occasion to talk on the phone with a high school senior, Noah Snyder, who was interested in math. I told him about the abc conjecture, and the fact that it was proved for polynomials. I suggested to him to think about the problem, and to try figure out a proof for polynomials. Remarkably enough, he found his own proof which is simpler than the one I learned from Mason. I am much indebted to Noah Snyder for letting his proof be published here. He is now an undergraduate at Harvard.   S.L.*]

Another proof for polynomial *abc* by Noah Snyder

We need the following lemma about polynomials, say with coefficients in the complex numbers. We denote the gcd of two polynomials f, g by

$$\gcd(f, g) = (f, g).$$

We let $n_0(f)$ be the number of distinct roots of the polynomial f.

Lemma. *Let f be a non-zero polynomial. Then*

$$\deg f = \deg(f, f') + n_0(f).$$

Proof. First a remark about a root of a polynomial f. Let α be a root of f, that is a complex number such that $f(\alpha) = 0$. We suppose f is not a constant polynomial so f has degree $d > 0$. We can write $f(t)$ as a polynomial in $(t - \alpha)$, that is a sum of descending powers of $(t - \alpha)$,

$$f(t) = c_d(t - \alpha)^d + \cdots + c_e(t - \alpha)^e,$$

with coefficients $c_d, \ldots, c_e$, and $e \geq 0$. Since α is a root, we actually have $e \geq 1$ and $c_e \neq 0$. Then

$$f'(t) = dc_d(t - \alpha)^{d-1} + \cdots + ec_e(t - \alpha)^{e-1},$$

and $ec_e \neq 0$. Hence we find that $(t - \alpha)^{e-1}$ is the largest power of $(t - \alpha)$ dividing $f'(t)$. Now let $\alpha_1, \ldots, \alpha_r$ be the distinct roots of f, and write

$$f(t) = c(t - \alpha_1)^{e_1} \cdots (t - \alpha_r)^{e_r} \quad \text{with some constant} \quad c \neq 0.$$

Then the only factors of degree 1 of (f, f') have to be $(t - \alpha_j)$ for some j, that is, for some non-zero constant c',

$$(f, f') = c'(t - \alpha_1)^{k_1} \cdots (t - \alpha_r)^{k_r}.$$

By unique factorization, the first remark shows that $k_j = e_j - 1$ for each $j = 1, \ldots, r$. Then we get

$$\begin{aligned} \deg f = e_1 + \cdots + e_r &= (e_1 - 1) + \cdots + (e_r - 1) + r \\ &= \deg(f, f') + n_0(f), \end{aligned}$$

which proves the lemma.

Theorem. *Let f, g, h be relatively prime polynomials such that $f + g = h$. Without loss of generality, let h be the polynomial of largest degree among f, g, h. Then*

$$\deg(h) \leqq n_0(fgh) - 1.$$

Proof. We are given $f + g = h$. Therefore $f' + g' = h'$. So

$$f'g - fg' = f'(f + g) - f(f' + g') = f'h - fh'.$$

We notice that (f, f') divides the left side, (g, g') divides the left side, and (h, h') divides the right side. Therefore, since f, g, h are relatively prime,

$$(f, f')(g, g')(h, h') \text{ divides } f'g - fg'.$$

So,

$$\deg(f, f') + \deg(g, g') + \deg(h, h') \leqq \deg(f'g - fg') \leqq \deg f + \deg g - 1.$$

Adding $\deg h$ to both sides and rearranging terms, we find that

$$\deg h \leqq \deg f - \deg(f, f') + \deg g - \deg(g, g') + \deg h - \deg(h, h') - 1.$$

By the lemma, we get

$$\deg h \leqq n_0(f) + n_0(g) + n_0(h) - 1 = n_0(fgh) - 1$$

since f, g, h are relatively prime.   Q.E.D.

Bibliography

[BLSTW 83] J. BRILLHART, D. H. LEHMER, J. L. SELFRIDGE, B. TUCKERMAN, and S. S. WAGSTAFF, Factorization of $b^n \pm 1, b = 2, 3, 5, 6, 7, 10, 11$ up to high powers, *Contemporary Math.*, Vol. 22, AMS, 1983

[Da 65] H. DAVENPORT, On $f^3(t) - g^2(t)$, *K. Norske Vid. Selsk. Forrh.* (Trondheim) **38** (1965) pp. 86-87

[Ha 71] M. HALL, The diophantine equation $x^3 - y^2 = k$, *Computers in Number Theory* (A. O. L. Atkin and B. J. Birch, eds.), Academic Press, 1971 pp. 173–198

[La 87] S. LANG, *Undergraduate Algebra*, Springer-Verlag, 1987, 1990. See especially Chapter IV, §9.

[La 90] S. LANG, Old and new conjectured diophantine inequalities, *Bull. AMS* **23** (1990) pp. 37–75

[Ma 83] R. C. MASON, *Diophantine Equations over Function Fields*, London Math. Soc. Lecture Note Series, Vol. 96, Cambridge University Press, 1984

[StT 86] C. L. STEWART and R. TIJDEMAN, On the Oesterle–Masser conjecture, *Monatshefte Math.* **102** (1986) pp. 251–257

[St 81] W. STOTHERS, Polynomial identities and hauptmoduln, *Quart. Math. Oxford* (2) **32** (1981) pp. 349–370

[Wi 95] A. WILES, Modular elliptic curves and Fermat's Last Theorem, *Annals of Math.* **142** (1995) pp. 443–551

[Za 95] U. ZANNIER, On Davenport's bound for the degree of $f^3 - g^2$ and Riemann's existence theorem, *Acta Arith.* **LXXI.2** (1995) pp. 107–137

Global Integration of Locally Integrable Vector Fields

The material of this talk stems from Artin's presentation of Cauchy's theorem, but really belongs in connection with basic real analysis, in a course on calculus of several (actually two) variables. No knowledge of complex analysis is assumed. Some basic knowledge about partial derivatives, is taken for granted, but no more than what is in [La 87], Chapter VIII, §4. A more complete treatment will be found in [La 97], Chapters XV and XVI.

Fundamental use is made of rectangular paths. If you are willing to accept a priori the notion of winding number for a closed rectangular path, then you can read Theorem 5 and its proof independently of everything else. The result is at the level of plane geometry technically, although conceptually things are put together in a somewhat more sophisticated way. It's really a result about closed circuits in the plane. When can a closed circuit be expressed as a sum of the boundaries of rectangles?

Global Integration of Locally Integrable Vector Fields

I'm going to deal with a topic which has an analogue usually covered in courses on complex analysis, but which is basically concerned with real analysis in the plane $\mathbf{R}^2$. Let U be a connected open set in $\mathbf{R}^2$. Consider a vector field F on U. This means that F is a mapping

$$F: U \to \mathbf{R}^2$$

given by coordinate functions

$$F(x, y) = (p(x, y), q(x, y)).$$

We suppose throughout that the component functions p, q have continuous partial derivatives, so as one says, are of class C^1. Let γ be a piecewise C^1 path in U. **Paths** *will always mean parametrized paths, so a C^1 curve in U is a mapping*

$$\gamma: [a, b] \to U$$

with continuous derivatives. A piecewise C^1 path is a sequence of C^1 curves

$$\gamma_i : [a_i, b_i] \to U, \qquad i = 1, \ldots, r,$$

such that $\gamma_i(b_i) = \gamma_{i+1}(a_i)$ for $i = 1, \ldots, r - 1$. In other words, the end point of the i-th map is the beginning point of the $(i + 1)$th map. So a path goes from the beginning point $\gamma_1(a_1)$ to the end point $\gamma_r(b_r)$. Of course, one could reparametrize the whole path so it is defined on a single interval $\gamma : [a, b] \to U$, with $b_i = a_{i+1}$ ($i = 1, \ldots, r - 1$).

Given our vector field F, we can define the integral of F along the path

$$\int_\gamma F = \sum_{i=1}^{r} \int_i F.$$

Each term is defined in the usual way,

$$\int_{\gamma_i} F = \int_{a_i}^{b_i} F(\gamma_i(t))\gamma_i'(t)\,dt.$$

The integral of F over γ in general depends on the path, and not just on the end points.

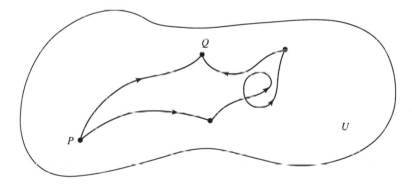

The situation is simpler for an important class of vector fields. A function $\varphi : U \to \mathbf{R}$ is called a **potential**, or a **potential function**, of the vector field F, if F is the gradient of φ, that is $F = \operatorname{grad} \varphi$. In other words,

$$(p, q) = \operatorname{grad} \varphi = \left(\frac{\partial \varphi}{\partial x}, \frac{\partial \varphi}{\partial y} \right).$$

A vector field which has a potential is called a **conservative** vector field. It's a simple theorem, standard in courses, that *if F is conservative, if φ is a potential function, and P, Q are the beginning and end points of a path γ, then*

$$\int_\gamma F = \varphi(0) - \varphi(P).$$

Thus in this case, the integral is independent of the path between the end points. The converse is also true. Precisely stated:

Theorem 1. *Let F be a vector field on the open connected set U in $\mathbf{R}^2$. Then F has a potential function on U if and only if given two points P, Q in U, and any path γ from P to Q in U, the integral*

$$\int_\gamma F$$

does not depend on γ.

Theorem 1 is a fundamental result, routinely proved in standard analysis courses. The idea is that one fixes a point P_0, and one defines

$$\varphi(Q) = \int_{P_0}^Q F,$$

along any path from P_0 to Q. Assuming that the integral is independent of the path, this integral gives a well-defined function, for which it is easy to prove that grad $\varphi = F$. We don't reproduce the proof here. Cf. for instance [La 83/97], Chapter XV, Theorem 4.2; or [La 87], Chapter VIII, §4.

It is also easy to prove that two potential functions on U differ by a constant, that is the difference is constant. Cf. [La 87], Chapter VII, Theorem 1.1. So a potential function is unique, up to an additive constant. Observe that this "constant of integration" disappears when we take the difference $\varphi(Q) - \varphi(P)$, giving the value of the integral of F when F has a potential function φ.

One can express Theorem 1 in another way. A path is called *closed* if its end point is the same as its beginning point.

Theorem 1'. *Let U be open connected in $\mathbf{R}^2$ and let F be a vector field on U. Then F has a potential function if and only if for all closed paths γ in U we have*

$$\int_\gamma F = 0.$$

To prove the equivalence between Theorem 1' and Theorem 1, one notes that following a path from P to Q with a path from Q to P yields a closed path. The argument is simple, and done routinely in courses.

Now suppose the vector field is of class C^2, in other words, its coordinate functions $p(x, y)$ and $q(x, y)$ have continuous partial derivatives up to order 2. Suppose F has a potential function φ, so by definition

$$p(x, y) = \frac{\partial \varphi}{\partial x} \quad \text{and} \quad q(x, y) = \frac{\partial \varphi}{\partial y}.$$

It is an elementary fact of analysis that

(2)
$$\frac{\partial p}{\partial y} = \frac{\partial q}{\partial x},$$

because the partials commute, that is

$$\frac{\partial p}{\partial y} = \frac{\partial^2 \varphi}{\partial y\, \partial x} = \frac{\partial^2 \varphi}{\partial x\, \partial y} = \frac{\partial q}{\partial x}.$$

Thus condition (2) is a necessary condition for F to have a potential function. This condition is called the **integrability condition**. In elementary analysis courses, it is routinely proved that locally, the integrability condition is sufficient for the existence of a potential function. More precisely:

Theorem 2. *Let F be a C^2 vector field on U satisfying the integrability condition (2). Let R be a rectangle contained in U, resp. a disc contained in U. Then F has a potential function φ on R, resp. D.*

Proof. Fix a point (x_0, y_0) in R, say the center of the rectangle, or in the case of the disc, let (x_0, y_0) be the center of the disc. For a point (x, y) in R, resp. D, define

$$\varphi(x, y) = \int_{x_0}^{x} p(t, y)\, dt + \int_{y_0}^{y} q(x_0, u)\, du.$$

The second integral is independent of x. By the fundamental theorem of calculus, we then get

$$\frac{\partial \varphi}{\partial x} = p(x, y),$$

which is the first of two conditions for a potential function. To find the partial $\partial \varphi / \partial y$ we have to differentiate under the integral sign, which you should know from an elementary analysis course. Then we find

$$\frac{\partial}{\partial y}\varphi(x, y) = \int_{x_0}^{x} \frac{\partial}{\partial y}\varphi(t, y)\, dt + q(x_0, y)$$

$$= \int_{x_0}^{x} \frac{\partial}{\partial t}q(t, y)\, dt + q(x_0, y) \quad \text{[by (2)]}$$

$$= q(x, y) - q(x_0, y) + q(x_0, y)$$

$$= q(x, y).$$

This proves the theorem.

A vector field which satisfies the integrability condition (2) will be called **locally integrable**, in light of Theorem 2. One could weaken this definition, by requiring only that given a point, there exists a neighborhood of the point on which the vector field has a potential function, but for simplicity, we stick to the definition we have given.

The whole point of our present discussion is to investigate conditions under which F has a potential function over all of U. In other words, we want global integrability. In general, the integrability condition is not sufficient. There is a standard example, namely the vector field given by

$$G(x, y) = \left(\frac{-y}{x^2 + y^2}, \frac{x}{x^2 + y^2}\right) = \left(\frac{-y}{r^2}, \frac{x}{r^2}\right) \qquad \text{where} \quad r^2 = x^2 + y^2.$$

Of course, this vector field is not defined at the origin $(0, 0) = O$. It is defined on the punctured plane $\mathbf{R}^2 - \{O\}$. An easy computation shows that $G(x, y)$ satisfies the integrability condition (2), and so G is locally integrable. However, G does not have a potential function on the whole domain of definition. To see this, we shall exhibit a closed curve γ such that

$$\int_{\gamma} G \neq 0.$$

In fact, we can take for γ any circle centered at the origin, as we shall now show.

First, we determine $p\, dx + q\, dy$ in polar coordinates, and it turns out that for $p = -y/r^2, q = x/r^2$ we have

$$p\, dx + q\, dy = d\theta.$$

Indeed, put $x = r \cos \theta$ and $y = r \sin \theta$ in polar coordinates. Then

$$dx = \frac{\partial x}{\partial r} dr + \frac{\partial x}{\partial \theta} d\theta = \cos \theta \, dr - r \sin \theta \, d\theta,$$

$$dy = \frac{\partial y}{\partial r} dr + \frac{\partial y}{\partial \theta} d\theta = \sin \theta \, dr - r \cos \theta \, d\theta.$$

Simple cancellations show that

$$(-y/r^2) \, dx + (x/r^2) \, dy = d\theta.$$

From this computation, it follows that the polar angle θ is a local potential function for G. If R is a rectangle in $\mathbf{R}^2 - \{O\}$, or D is a disc in $\mathbf{R}^2 - \{O\}$, then θ is a potential for G in R, resp. D.

Let γ be a path which does not contain the origin O. Then

$$\int_\gamma G = \int_\gamma d\theta.$$

If γ is a closed path in the punctured plane $\mathbf{R}^2 - \{O\}$, then along the path, θ changes from the beginning point to the end point by an integral multiple of 2π. Thus

(3)
$$\int_\gamma G = 2\pi k,$$

where k is an integer. In particular, for a circle centered at the origin, oriented counterclockwise, say the unit circle $x = \cos \theta$, $y = \sin \theta$, we find

$$G = \int_0^{2\pi} d\theta = 2\pi \neq 0.$$

As you can verify at once, the same holds for any circle centered at the origin. In particular, G does not have a potential on $\mathbf{R}^2 - \{O\}$.

The relation (3) is fundamental, and is the basis for the following definition. Let γ be a closed path in the plane, not containing the origin O. We define the **winding number** of the path

$$W(\gamma, O) = \frac{1}{2\pi} \int_\gamma G.$$

This is the winding number with respect to the origin. For an arbitrary point P, we can also define the winding number by making a translation.

Let $P = (x_0, y_0)$. Let G_P be the vector field defined by

$$G_P(x, y) = G(x - x_0, y - y_0).$$

Let γ be a closed path not containing P. We define the winding number $W(\gamma, P)$ by the formula

$$W(\gamma, P) = \frac{1}{2\pi} \int_\gamma G_P.$$

We shall see later that G and its translations G_P constitute the only obstructions for a locally integrable vector field to have a global potential function. This fundamental result is based on the next theorem, which will be shown to amount to a purely topological result, but which we first express analytically.

Theorem 3. *Let U be an open connected set in $\mathbf{R}^2$. Let γ be a closed path in U. Assume that for all points $P \notin U$ the winding number $W(\gamma, P)$ is equal to 0. Let F be a locally integrable vector field on U. Then*

$$\int_\gamma F = 0.$$

Actually, instead of dealing with paths in the strict sense we have defined them, it is convenient to deal with formal linear combinations

$$\gamma = \sum_{i=1}^n m_i \gamma_i$$

where each γ_i is a closed path and m_i is an integer for each i. The integral of F over γ is defined to be simply the sum of the integrals,

$$\int_\gamma F = \sum_{i=1}^n m_i \int_{\gamma_i} F.$$

Such a formal linear combination will be called a **closed chain**. Then both the theorem and lemma are valid for closed chains.

Example. We could let γ_1 be the circle of radius 1 centered at the origin, and γ_2 the circle of radius 2. Let

$$\gamma = \gamma_1 - \gamma_2.$$

Each circle is oriented counterclockwise. Take $F = G$, the vector field defined before Theorem 3, and let $U = \mathbf{R}^2 - \{O\}$ be the punctured plane. The only point not in U is the origin, and then the winding number $W(\gamma, O)$ turns out to be

$$2\pi W(\gamma, O) = \int_{\gamma_1} G - \int_{\gamma_2} G = 2\pi - 2\pi = 0.$$

We shall now give the proof of Theorem 3. Then we shall give applications.

Proof of the main theorem

The proof will take several steps. The first step is to reduce the theorem to what we call rectangular paths. Precisely, we define a path to be **rectangular** when it consists of a finite number of line segments which are parallel to the coordinate axes, so line segments which are horizontal or vertical. We then have the reduction lemma:

Lemma 4. *If Theorem 3 is true for rectangular closed chains, then it is true for arbitrary closed chains (always assumed piecewise continuously differentiable).*

To prove this lemma, we need to show first how an arbitrary path γ in U can be replaced by a rectangular path η in U having the same beginning point and the same end point, and such that for every locally integrable vector field F in U, we have

$$\int_\eta F = \int_\gamma F.$$

Suppose that $\gamma \colon [a, b] \to U$ is defined over a closed interval $[a, b]$. By the uniform continuity of γ, one can use a partition of the interval

$$a = a_1 \leqq a_2 \leqq \cdots \leqq a_N = b$$

with $a_{i+1} - a_i$ sufficiently small such that the image of $[a_i, a_{i+1}]$ under γ is contained in a disc $D_i \subset U$. We let $\gamma_i \colon [a_i, a_{i+1}] \to U$ be the restriction of γ to the small interval $[a_i, a_{i+1}]$, and we let $P_i = \gamma(a_i)$.

Now we make up a rectangular path by replacing each γ_i with two line segments parallel to the axes, as shown on the figure. Thus η_i is also contained in D_i, and $\eta_1, \ldots, \eta_{N-1}$ constitute a rectangular path η from P_1 to P_N. Furthermore, for each i, the path η_i is also contained in D_i. Since

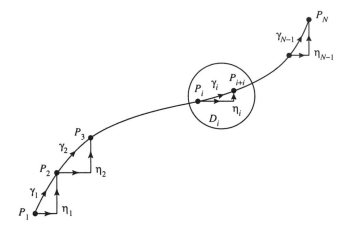

P_i, P_{i+1} are the beginning point and end point of both γ_i and η_i, and since F is locally integrable, it follows that

$$\int_{\gamma_i} F = \int_{\eta_i} F \qquad \text{for} \quad i = 1, \dots, N - 1.$$

Therefore

$$\int_{\gamma} F = \int_{\eta} F.$$

In particular, this applies to the vector field G, and therefore

$$\int_{\gamma} G = \int_{\eta} G.$$

By definition, this means for the winding numbers that

$$W(\eta, P) = W(\gamma, P) = 0$$

for every point P not in U. So if Theorem 3 is true for the rectangular path η, then it is true for γ, and Lemma 4 is proved.

There remains to prove Theorem 3 for rectangular paths or chains.

Suppose first that R is a closed rectangle contained in U. Let ∂R be its boundary, oriented counterclockwise as on the figure.

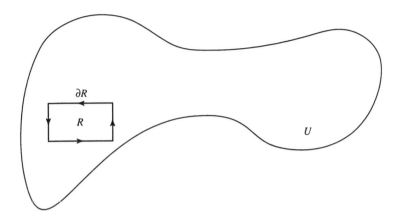

Since the vector field F is locally integrable, it follows from Theorem 2 that

$$\int_{\partial R} F = 0.$$

If a path, or chain is given as a sum of boundaries of rectangles, say

$$\eta = \sum_i m_i \, \partial R_i,$$

and each rectangle is contained in U, then

$$\int_\eta F - \sum_i m_i \int_{\partial R_i} F = 0.$$

The fundamental result is contained in the next theorem, and achieves what we want for closed rectangular chains. By Lemma 4, it also proves Theorem 3.

Rectangular closed paths
A theorem on circuits in the plane

Let U be our connected open set in $\mathbf{R}^2$ and let η be a rectangular path in U. Then η consists of successive horizontal and vertical segments. Let σ be such a segment, say $\sigma \colon [a, b] \to U$. Pick $a < c < b$. Then σ is subdivided into two segments $\sigma' \colon [a, c] \to U$ and $\sigma'' \colon [c, b] \to U$.

$$\sigma(a) \quad \sigma(c) \qquad \sigma(b)$$
$$\bullet\underset{\sigma'}{\rule{2cm}{0.4pt}}\bullet\underset{\sigma''}{\rule{2.5cm}{0.4pt}}\bullet$$

If we replace σ by $\sigma' + \sigma''$, and iterate this procedure a finite number of times, applied to the various segments constituting η, we obtain what we shall call a **subdivision** of η, denoted by Subd η. We can now state a fundamental theorem in which only the winding number appears, but otherwise no vector field.

Theorem 5. *Let η be a closed rectangular chain in U. Assume that $W(\eta, P) = 0$ for every point P not in U. Then there exist a finite number of rectangles R_i $(i = 1, \ldots, N)$ contained in U, and integers m_i, such that for some subdivision of η, we have*

$$\text{Subd } \eta = \sum_i m_i \, \partial R_i.$$

Proof. Given the rectangular chain η, we draw all vertical and horizontal lines passing through the sides of the chain, as illustrated on the next figure.

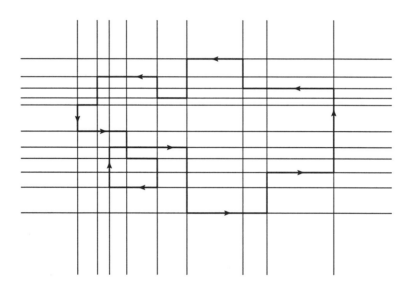

Then these vertical and horizontal lines decompose the plane into rectangles, and rectangular regions extending to infinity in the vertical and horizontal direction. Let R_i be one of the rectangles, and let P_i be a point inside R_i. Let

$$m_i = W(\eta, P_i).$$

For some rectangles we have $m_i = 0$, and for some rectangles, we have $m_i \neq 0$. We let $R_1, \ldots, R_N$ be those rectangles such that $m_1, \ldots, m_N$ are not 0, and we let ∂R_i be the boundary of R_i for $i = 1, \ldots, N$, oriented counterclockwise. We shall prove the following two assertions:

1. Every rectangle R_i such that $m_i \neq 0$ is contained in U.
2. Some subdivision of η is equal to

$$\sum_{i=1}^{N} m_i \, \partial R_i.$$

This will prove the desired theorem.

Assertion 1. By assumption, P_i must be in U, because $W(\eta, P) = 0$ for every point P outside U. Note that the winding number function

$$P \mapsto W(\eta, P)$$

is continuous, so constant on connected sets because it is integer valued. Hence for all P in the interior of R_i we have $W(\eta, P) = W(\eta, P_i) \neq 0$. Hence the interior of R_i is contained in U. If a boundary point of R_i is on η, then it is in U. If a boundary point of R_i is not on η, then the winding number with respect to η is defined, and again is $\neq 0$ by continuity, approaching the boundary point from the interior. This proves that the whole rectangle, including the boundary, is contained in U, and proves our first assertion.

Assertion 2. To prove this assertion, we first observe that if P_k is a point outside the rectangle R_i, then

$$W(\partial R_i, P_k) = 0.$$

However, for Q inside R_i, we have $W(\partial R_i, Q) = 1$.

Now we come to the proof proper for the assertion. All the vertical and horizontal lines passing through our chain η may cut some of the horizontal and vertical paths occurring in η into several pieces, thus giving rise to a subdivision of η. For purposes of integration, this subdivision amounts to η. Now we consider the chain

$$C = \eta - \sum_{i} m_i \, \partial R_i.$$

Assertion 2 amounts to showing that the chain C does not contain any segment, in other words, that the chain C is the zero chain. Suppose some segment σ occurs in C, and σ is the side of a rectangle R_k. Then we can write

$$C = \eta - \sum_{i} m_i \, \partial R_i = m\sigma + \text{terms without } \sigma,$$

where m is some integer, which is $\neq 0$ if σ occurs in C. Actually, we shall prove that $m = 0$, so no segment can remain in C. We consider the chain

$$C' = \eta - \sum m_i \, \partial R_i - m \, \partial R_k = C - m \, \partial R_k,$$

and we compute the winding number with respect to P_k which lies inside R_k. We find

$$
W\left(\eta - \sum_i m_i \, \partial R_i - m \, \partial R_k, \, P_k \right)
$$

$$
= W(\eta, P_k) - \sum_i m_i \, W(\partial R_i, \, P_k) - m W(\partial R_k, \, P_k)
$$

$$
= m_k - m_k - m
$$

$$
= - m.
$$

However, the chain C' does not contain σ any more, because σ is a side of R_k and we subtracted $m\sigma$ from C by subtracting $m \, \partial R_k$ from C. In our decomposition of the plane into rectangles, σ belongs to the boundary of two rectangles. If R_k lies on one side of σ, then some rectangle R_j lies on the other side, as on the next figure.

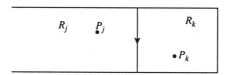

The points P_k and P_j can be joined by a line segment which does not have any point in common with the chain C'. Hence

$$
-m = W\left(\eta - \sum m_i \, \partial R_i - m \, \partial R_k, \, P_k \right)
$$

$$
= W\left(\eta - \sum m_i \, \partial R_i - m \, \partial R_k, \, P_j \right)
$$

$$
= W(\eta, P_j) - \sum_i m_i W(\partial R_i, P_j) - m W(\partial R_k, P_j)
$$

$$
= m_j - m_j - 0
$$

$$
= 0.
$$

This proves Assertion 2, and concludes the proof of Theorem 5.

Remark. The above proof contains no analysis, other than the definition of the winding number, which it was convenient to give as an integral.

Otherwise, the arguments are purely combinatorial, and the theorem is a result concerning circuits in the plane.

Two applications

We shall apply Theorem 3 in two ways. The first will show how the integral along a closed path or chain can be evaluated as a sum of integrals over small circles. The second will give a condition showing that the only obstruction to a locally integrable vector field having a potential function is due to our vector field G and its translates G_P over certain points P.

Theorem 6. *Let U be an open set in $\mathbf{R}^2$ and let γ be a closed chain in U such that $W(\gamma, P) = 0$ for all $P \notin U$. Let $P_1, \ldots, P_n$ be a finite number of distinct points of U not on γ. Let γ_i $(i = 1, \ldots, n)$ be the boundary of a closed disc $\bar{D}_i$ contained in U, containing P_i, and oriented counterclockwise. We assume that $\bar{D}_i$ does not intersect γ and does not intersect $\bar{D}_j$ if $i \neq j$. Let*

$$m_i = W(\gamma, P_i).$$

Let U^* be the set obtained by deleting $P_1, \ldots, P_n$ from U. Let F be a locally integrable vector field on U^*. Then

$$\int_\gamma F = \sum_{i=1}^n m_i \int_{\gamma_i} F.$$

Proof. Let

$$C = \gamma - \sum m_i \gamma_i.$$

We claim that for all points P outside U^* we have $W(C, P) = 0$. If $P \notin U$, then

$$W(C, P) = W(\gamma, P) - \sum m_i W(\gamma_i, P) = 0 - 0 = 0.$$

Next, for one of the points P_k instead of P, we have

$$W(\gamma_i, P_k) = \begin{cases} 1 & \text{if } i = k, \\ 0 & \text{if } i \neq k. \end{cases}$$

Therefore

$$W(C, P_k) = W(\gamma, P_k) - \sum m_i W(\gamma_i, P_k) = W(\gamma, P_k) - m_k = 0.$$

We then apply Theorem 3 to the open set U^* and to the chain C. We obtain

$$0 = \int_C F = \int_\gamma F - \sum_{i=1}^n m_i \int_{\gamma_i} F,$$

which proves Theorem 6.

Example 1. In the next figure, we have

$$\int_\gamma F = -\int_{\gamma_1} F - 2\int_{\gamma_2} F - \int_{\gamma_3} F - 2\int_{\gamma_4} F.$$

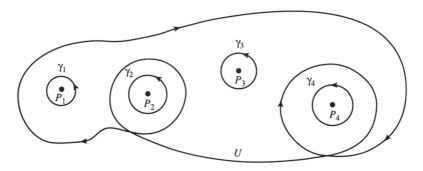

Let P be a point of $\mathbf{R}^2$ and let D_P be a disc centered at P. Let D_P^* be the punctured disc obtained by deleting P. Let γ_P be a circle centered at P in the disc D_P, and oriented counterclockwise. Let F be a locally integrable vector field on D_P^*. We define the **residue** of F at P to be

$$\operatorname{res}_P(F) = \frac{1}{2\pi} \int_{\gamma_P} F.$$

This residue is independent of the choice of circle γ_P, as we see by invoking Theorem 6. Applying Theorem 6 then yields the formula

$$\boxed{\int_\gamma F = \sum_i 2\pi m_i \cdot \operatorname{res}_{P_i}(F).}$$

In other words, the integral of F along γ is equal to 2π times the sum of the residues, counted with the multiplicities m_i.

This terminology is used in complex function theory, where you will find the theorem corresponding to Theorem 6 formulated for certain complex functions. But when you take a course on that subject, you should verify for yourself that Theorem 6 which we have proved gives the corresponding theorem about complex functions as a special case. This theorem is called the **residue theorem**, or **Cauchy's residue theorem**.

We come to the second application, giving the obstruction to the existence of a potential function.

Theorem 7. *Let U be a connected open set in $\mathbf{R}^2$ such that for every closed path γ in U and every point $P \notin U$ we have*

$$W(\gamma, P) = 0.$$

Let $P_1, \ldots, P_n$ be distinct points of U, and let $U^* = U - \{P_1, \ldots, P_n\}$ be the set obtained by deleting $P_1, \ldots, P_n$ from U. Let F be a locally integrable vector field on U^*. As usual, let

$$G(x, y) = \left(\frac{-y}{x^2 + y^2}, \frac{x}{x^2 + y^2} \right).$$

Let G_P be its translation by a point P. Then there exist constants $a_1, \ldots, a_n$ and a function φ on U^* such that

$$F - \sum_{i=1}^{n} a_i G_{P_i} = \operatorname{grad} \varphi.$$

Proof. Let γ_i be a small circle centered at P_i, oriented counterclockwise. We assume the circles small enough that if $k \neq i$, then the closed discs bounded by γ_i and γ_k are disjoint. We then have:

$$(*) \qquad \int_{\gamma_i} G_{P_j} = \begin{cases} 2\pi & \text{if } i = j, \\ 0 & \text{if } i \neq j. \end{cases}$$

Let

$$a_i = \frac{1}{2\pi} \int_{\gamma_i} F = \operatorname{res}_{P_i}(F) \quad \text{in our latest terminology.}$$

Let

$$F_1 = F - \sum_{i=1}^{n} a_j G_{P_j}.$$

By Theorem 1' and Theorem 3, it suffices to prove that for all closed path

γ in U^* we have

$$\int_\gamma F_1 = 0.$$

But using Theorem 6, we find

$$\int_\gamma F_1 = \sum_i m_i \int_{\gamma_i} F_1 = \sum_i m_i \int_{\gamma_i} \left(F - \sum_j a_j G_{P_j} \right)$$

$$= \sum_i m_i 2\pi a_i - \sum_i \sum_j \int_{\gamma_i} m_i a_j G_{P_j}$$

$$= \sum_i m_i 2\pi a_i - \sum_i m_i a_i 2\pi \quad \text{[by using (*)]}$$

$$= 0.$$

This proves Theorem 7.

Bibliography

[Ar 65] E. ARTIN, On the theory of complex functions, *Collected Papers of Emil Artin*, Addison-Wesley, 1965; pp. 513–522. Reprinted by Springer-Verlag, Second edition, 1996

[La 87] S. LANG, *Calculus of Several Variables*, Springer-Verlag, 1987

[La 97] S. LANG, *Undergraduate Analysis*, Second edition, Springer-Verlag, 1997

[La 99] S. LANG, *Complex Analysis*, Fourth edition, Springer-Verlag, 1999

Approximation Theorems of Analysis

Analysis to a large extent makes estimates. But there is also a component to analysis which furnishes identities between various expressions, so an essentially algebraic structure as distinguished from estimates. Dirac sequences and Dirac families involve both aspects.

The examples show a broad range of applications. Aside from that, I found it natural to mention theta series in connection with completely different contexts, to show how seemingly different parts of mathematics are in fact closely related. Courses usually stick to one subject, because of time pressures, and because they are preparatory to other courses, so there isn't much time to look out the window and meander. In the present book, we have no fixed schedule, no syllabus, no tests, and the boundary conditions are completely different from those of a course. The material provides opportunities for a whole series of talks.

Approximation Theorems of Analysis

I'm going to start with a very general approximation theorem, which concerns so-called Dirac sequences. Then I'll discuss examples and applications, such as:

- The Weierstrass approximation theorem, which says that a continuous function on a finite closed interval can be uniformly approximated by polynomials.
- Fourier series.
- Harmonic functions and Poisson families.
- Solutions of the heat equation.

In each case, the particular result in a particular direction will be seen to be a special case of the general theorem.

Except for the Weierstrass approximation theorem, the items were originally discovered in areas common to math and physics. But just because something is discovered in connection with physics does not mean that it "is" physics. All of the above items have significance in many areas of mathematics, and possibly all areas of mathematics. I like to say, for instance, that the heat kernel which we'll discuss later is the big bang of the mathematical universe. You can find it all over the place.

Now to mathematics. We are on the real line **R**. We begin by defining the convolution of two functions. Let f, g be two functions on **R** (the real numbers), continuous, or piecewise continuous, complex valued, and such that for $x \to \pm\infty$, the values $f(x)$ tend to 0 sufficiently rapidly so that all the forthcoming integrals converge. For instance, if $f(x) = 0$ outside a finite interval, then f meets the above conditions. We define the **convolution** $f * g$ by the formula

$$f * g(x) = \int_{-\infty}^{\infty} f(x - t)g(t)\,dt.$$

One sees at once that convolution is commutative, that is

$$f * g = g * f.$$

We just make the substitution $u = x - t$ and $t = x - u$, $du = -dt$. We view convolution as a product, which is bilinear, that is

$$f * (g_1 + g_2) = f * g_1 + f * g_2 \qquad \text{and} \qquad (f_1 + f_2) * g = f_1 * g + f_2 * g;$$

furthermore, for every complex number α,

$$(\alpha f) * g = f * (\alpha g) = \alpha(f * g).$$

Also by interchanging the order of integration, one can verify associativity for three functions f, g, h, namely

$$(f * g) * h = f * (g * h).$$

Thus the convolution product satisfies the ordinary rules of multiplication. One says that the piecewise continuous functions with sufficiently fast decay at infinity form a commutative algebra over **C**, and the real valued functions form a commutative algebra over **R**.

There arises a first question: Is there a unit element in these algebras? In other words, is there a function δ such that $\delta * f = f$ for all functions f, say continuous functions vanishing outside some finite interval? Who says yes? (*Some hands are raised.*) Who says no? (*Again some hands are raised. Most keep silent.*) I'll give you the answer. The answer is NO. But we are going to define something which is almost as good as a unit element, namely a Dirac sequence.

By definition, a **Dirac sequence** is a sequence of continuous functions $\{K_n\}$ having the following properties:

DIR 1. The functions K_n are semipositive, that is $K_n \geq 0$ for all $n = 1, 2, \ldots$.

DIR 2. For all n, we have

$$\int_{-\infty}^{\infty} K_n = 1.$$

DIR 3. Given $\epsilon > 0$ and $\delta > 0$ there exists n_0 such that for all $n \geqq n_0$ we have the estimate

$$\int_{|x| \geqq \delta} K_n(x)\,dx < \epsilon.$$

A Dirac sequence may look as in the following figure:

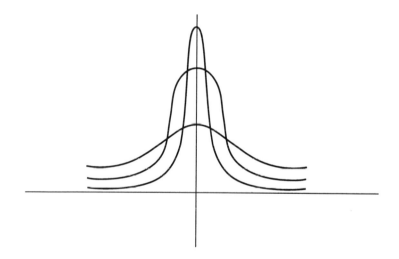

In practice, Dirac sequences consist of even functions, that is their graph is symmetric with respect to the vertical axis, as we have drawn on the figure.

Dirac sequences have a remarkable approximation property, which we state and prove.

Theorem 1 (General approximation theorem). *Let f be a bounded piecewise continuous function on $\mathbf{R}$. Then the sequence $\{K_n * f\}$ converges to f uniformly on every compact set where f is continuous. That is for each point x at which f is continuous, we have*

$$\lim_{n \to \infty} (K_n * f)(x) = f(x),$$

and the convergence is uniform as stated above.

In a certain sense, the sequence $\{K_n\}$ "converges" toward a unit element, even though it is clear from the picture that the sequence does not have a limit *function*. Because of the above limit property, one usually says that a Dirac sequence is an **approximation of the identity**.

Proof. The proof is remarkably simple. By definition,

$$K_n * f(x) = f * K_n(x) = \int_{-\infty}^{\infty} K_n(t) f(x - t) \, dt.$$

By property **DIR 2**, we have

$$f(x) = f(x) \int_{-\infty}^{\infty} K_n(t) \, dt = \int_{-\infty}^{\infty} K_n(t) f(x) \, dt.$$

Subtracting, we get

$$K_n * f(x) - f(x) = \int_{-\infty}^{\infty} K_n(t)[f(x - t) - f(x)] \, dt.$$

Let B be a bound for f on $\mathbf{R}$, that is

$$|f(x)| \leq B \qquad \text{for all } x \in \mathbf{R}.$$

Let S be a compact set where f is continuous. Then f is uniformly continuous on S, and that's the only property we shall use. So given ϵ, there exists δ such that for all $x \in S$ and $|t| \leq \delta$, we have

$$|f(x - t) - f(x)| < \epsilon.$$

Then we estimate the convolution integral, which we split into two pieces:

$$|K_n * f(x) - f(x)|$$
$$= \int_{|t| \leq \delta} K_n(t)|f(x - t) - f(x)| \, dt + \int_{|t| \geq \delta} K_n(t)|f(x - t) - f(x)| \, dt$$

$$\leq \int_{-\infty}^{\infty} K_n(t) \epsilon \, dt + \epsilon 2B \quad \text{for } n \geq n_0 \text{ as in } \textbf{DIR 3}$$

$$\leq \epsilon + \epsilon 2B.$$

This proves that $\{K_n * f\}$ approximates f uniformly on S, and concludes the proof of the theorem.

We shall list several specific applications of the general theorem to different areas of mathematics.

The Weierstrass approximation theorem

Theorem. *Let f be a continuous function on a finite closed real interval* $[a, b]$. *Then f can be uniformly approximated by polynomials on* $[a, b]$.

Proof. By a translation and a dilation, amounting to a change of variable, we may assume without loss of generality that the interval is $[0, 1]$. Let L be the linear function such that $L(0) = f(0)$ and $L(1) = f(1)$. Since L is a polynomial (of degree 1), it suffices to prove the theorem for the function $f - L$ instead of f. The advantage we have gained is that $f - L$ vanishes at the end points of the interval. Thus we are reduced to proving the theorem when $f(0) - f(1) = 0$, which we assume from now on.

We then define the **Landau sequence**

$$K_n(x) = \begin{cases} \frac{1}{c_n}(1 - x^2)^n & \text{if } -1 \leq x \leq 1, \\ 0 & \text{if } |x| \geq 1. \end{cases}$$

We choose the constant c_n so that

$$\int_{-1}^1 K_n(x)\, dx = 1 \qquad \text{so} \qquad c_n = \int_{-1}^1 (1 - x^2)^n\, dx.$$

Then K_n satisfies **DIR 2**. Trivially, $K_n(x) \geq 0$ for all x. It's easy to prove that **DIR 3** is satisfied. You can work out the details, or look them up in [La 83/97], Chapter XI, §2, pp. 288–289. Thus $\{K_n\}$ is a Dirac sequence. By the general approximation theorem, it follow that $K_n * f$ converges uniformly to f on the interval $[0, 1]$. All that remains to be proved is that $K_n * f$ is a polynomial. This is easy, and we carry out the computation. We have

$$K_n * f(x) = \int_{-1}^1 K_n(x - t) f(t)\, dt.$$

We can expand $K_n(x - t)$ as a polynomial in x and t, namely

$$K_n(x - t) = \sum a_{ij}^{(n)} x^i t^j \qquad \text{with coefficients } a_{ij}^{(n)} \in \mathbf{R}.$$

Therefore

$$K_n * f(x) = \sum a_{ij}^{(n)} x^i \int_{-1}^1 t^j f(t)\, dt = \sum_i \sum_j a_{ij}^{(n)} b_j x^i$$

where

$$b_j = \int_{-1}^1 t^j f(t)\, dt.$$

Thus $K_n * f(x)$ is a polynomial in the variable x and the theorem is proved.

Who has previously seen a proof of the Weierstrass theorem?

A STUDENT. We have seen a different proof in a course.

SERGE LANG. What kind of a proof?

THE STUDENT. With the Weierstrass–Stone theorem.

SERGE LANG. Well, the Weierstrass–Stone theorem is a perfectly good general theorem, quite useful in many contexts. But there is an essential difference between applying the Weierstrass–Stone theorem and the proof we gave above. The Weierstrass–Stone theorem gives no information about the approximating sequence, whereas using the Landau kernel exhibits such a sequence explicitly in terms of the original function f. The Weierstrass–Stone theorem for example doesn't give you information about the degrees of the approximating polynomials as a function of n, nor does it give you an estimate of the coefficients of the approximating polynomials as a function of n, and of course the original function f. The explicit construction does tell you rather precisely and explicitly how the approximation to f takes place. Using the Landau functions provides an explicit approximating sequence, and in this sense it is better than the Weierstrass–Stone type of proof if you want an effective, constructive result.

Fourier series

In this application, we consider periodic functions, of period 2π. Let f be such a function. Integrals will consequently be taken over the interval $[-\pi, \pi]$ instead of $(-\infty, \infty)$ in the definition of a Dirac sequence. The main approximation theorem remains the same, except for this change. Also sometimes we have a normalization factor $1/2\pi$.

For any integer n, let

$$\chi_n(x) = e^{inx}.$$

We consider the convolution of χ_n and a function f (continuous if you want, at worst piecewise continuous), namely

$$(\chi_n * f)(x) = \int_{-\pi}^{\pi} \chi_n(x - t) f(t) \, dt$$

$$= \int_{-\pi}^{\pi} e^{in(x-t)} f(t) \, dt$$

$$= e^{inx} \int_{-\pi}^{\pi} e^{-int} f(t) \, dt$$

$$= 2\pi c_n e^{inx}$$

where

$$c_n = \frac{1}{2\pi} \int_{-\pi}^{\pi} e^{-int} f(t) \, dt$$

is known as the n-th **Fourier coefficient** of f. The **Fourier series** of f is then defined to be

$$S_f(x) = \sum_{-\infty}^{\infty} c_n e^{inx}.$$

It may or may not converge, because we have not made any assumption on f beyond continuity assumptions. We define the **Dirichlet kernel** D_n as the finite sum

$$D_n = \frac{1}{2\pi} \sum_{k=-n}^{n} \chi_k \quad \text{or also} \quad D_n(x) = \frac{1}{2\pi} \sum_{k=-n}^{n} e^{ikx}.$$

Then $S_{f,n} = D_n * f$ is the n-th partial sum of the Fourier series, namely

$$S_{f,n}(x) = \sum_{k=-n}^{n} c_k e^{ikx}.$$

Unfortunately, $\{D_n\}$ is not a Dirac sequence, and as we have already stated, the partial sums $S_{f,n}$ need not converge, let alone to the function f. However, we are going to construct another sequence of trigonometric polynomials, which is a Dirac sequence and converges to the function f.

We define K_n to be the average of the Dirichlet kernels, that is

$$K_n = \frac{1}{n}(D_0 + D_1 + \cdots + D_{n-1}),$$

or in terms of x,

$$K_n(x) = \frac{1}{n}(D_0(x) + \cdots + D_{n-1}(x)).$$

It is an easy computation with trigonometric identities and finite geometric series to find that

$$K_n(x) = \frac{1}{2\pi n} \frac{\sin^2(nx/2)}{\sin^2(x/2)}.$$

Having the square term on the right side shows that $K_n \geq 0$, which is the first condition **DIR 1** for a Dirac sequence.

To verify **DIR 2**, we first integrate $D_n(x)$ term by term. We have

$$\int_{-\pi}^{\pi} \chi_k(x)\,dx = \int_{-\pi}^{\pi} e^{ikx}\,dx = \begin{cases} 2\pi & \text{if } k = 0, \\ 0 & \text{if } k \neq 0. \end{cases}$$

Then we get at once

$$\int_{-\pi}^{\pi} D_n(x)\,dx = 1,$$

after integrating each term and taking the sum. This proves that the sequence $\{D_n\}$ satisfies **DIR 2**. But then taking the sum defining K_n from $D_0, \ldots, D_{n-1}$ we conclude that

$$\int_{-\pi}^{\pi} K_n(x)\,dx = 1,$$

which is **DIR 2** for K_n.

Finally, we want to verify **DIR 3**. For $0 < \delta \leq |x| \leq \pi$, the values

$$\left| \frac{1}{\sin^2(x/2)} \right|$$

are bounded, because $x/2$ stays away from 0. Let B be a bound, that is

$$\left| \frac{1}{\sin^2(x/2)} \right| \leq B.$$

Given ϵ there exists n_0 such that for $n \geq n_0$ we get

$$\frac{1}{2\pi n} \int_{\delta \leq |x| \leq \pi} \left| \frac{\sin^2(nx/2)}{\sin^2(x/2)} \right| dx \leq \frac{1}{n} B < \epsilon.$$

This proves **DIR 3**, and thus shows:

The sequence $\{K_n\}$ is a Dirac sequence (for periodic functions).

Applying the general approximation theorem, we get one of the basic results from the theory of Fourier series.

Theorem of Fejer–Cesaro. *Let f be a piecewise continuous periodic function. Let $\{K_n\}$ be the average of the partial sums of the Fourier series* *for f. Then $K_n(x)$ converges uniformly to $f(x)$ for all x in a compact set where f is continuous, that is*

$$\frac{1}{n}(S_0(x) + S_1(x) + \cdots + S_n(x)) \to f(x)$$

uniformly as stated above.

The summation of the averages of the partial sums of the Fourier series is known as **Cesaro summation**.

Harmonic functions on the disc

Our next examples will deal with a variation of the notion of sequence, namely we deal with a family rather than a sequence. Thus we shall define a Dirac family $\{K_r\}$ with $0 \leq r < 1$, and we let r tend to 1 rather than n tend to infinity. We call $\{K_r\}$ a **Dirac family** if it satisfies the three conditions:

DIR 1. For all r with $0 \leq r < 1$, we have $K_r(t) \geq 0$.

DIR 2. For these values of r, we have

$$\int_{-\pi}^{\pi} K_r(t)\,dt = 1.$$

DIR 3. Given ϵ, δ there exists r_0 with $0 < r_0 < 1$ such that for all r with $r_0 \leq r < 1$, we have

$$\int_{-\pi}^{-\delta} K_r(t)\,dt + \int_{\delta}^{\pi} K_r(t)\,dt < \epsilon.$$

In exactly the same way that we proved the first approximation theorem for sequences, one can prove the corresponding version for families.

Approximation theorem for Dirac families. *Let $\{K_r\}$ be a Dirac family as above. Let f be a piecewise continuous function. Then $K_r * f$ converges to f for $r \to 1$, uniformly on every compact set where f is continuous.*

Of course, we now want significant examples of Dirac families. We use polar coordinates (r, θ), and we define

$$P_r(\theta) = P(r, \theta) = \frac{1}{2\pi} \sum_{-\infty}^{\infty} r^{|n|} e^{in\theta} \qquad \text{for} \quad 0 \leq r < 1.$$

We shall prove that $\{P_r\}$ is a Dirac family, called the **Poisson family**. First observe that the series defining $P_r(\theta)$ is absolutely convergent in our range $0 \leq r < 1$, uniformly in an interval $0 \leq r \leq r_0$ if $r_0 < 1$, because the series can be compared with the geometric series. Simple trigonometric identities show that $P(r, \theta)$ has the expression

$$P(r, \theta) = \frac{1}{2\pi} \frac{1 - r^2}{1 - 2r\cos\theta + r^2}.$$

From this identity we can verify **DIR 1**, because both the numerator and denominator in the above expression are ≥ 0. In fact, the smallest value for the denominator is when $\cos\theta = 1$, in which case the denominator is equal to $(1 - r)^2$.

Next we show **DIR 2**. We integrate the series for $P(r, \theta)$ term by term, and use the same values that we found when we looked at Fourier series, namely

$$\int_{-\pi}^{\pi} e^{in\theta} d\theta = \begin{cases} 2\pi & \text{if } n = 0, \\ 0 & \text{if } n \neq 0. \end{cases}$$

It follows that only the integral of the term with $n = 0$ gives a non-zero contribution to the integral of the series, and we find

$$\int_{-\pi}^{\pi} P_r(\theta) d\theta = 1.$$

This proves **DIR 2**.

Finally we show that $\{P_r\}$ satisfies **DIR 3**. For $|\theta| \geq \delta > 0$, we have the inequality

$$\frac{1}{1 - 2r\cos\theta + r^2} \leq \frac{1}{1 - 2r\cos\delta + r^2}.$$

Using the derivative test on $1 - 2r\cos\delta + r^2$, we see at once that the minimum of the denominator on the right occurs when $r = \cos\delta$. At this minimum, the value of the denominator is $1 - (\cos\delta)^2$. Hence the quotient

$$\frac{1}{2 - 2r\cos\theta + r^2}$$

as function of r is uniformly bounded. Hence

$$\lim_{r \to 1} \frac{1 - r^2}{1 - 2r\cos\theta + r^2} = 0,$$

and the limit holds uniformly for $|\theta| \geq \delta > 0$. Hence there exists r_0 such that for $r_0 \leq r < 1$ we have

$$\int_{\delta \leq |\theta| \leq \pi} P_r(\theta) \, d\theta < \epsilon.$$

This proves that the family $\{P_r\}$ satisfies **DIR 3**, and hence is a Dirac family.

As a consequence of the general approximation theorem, we now obtain:

Poisson approximation theorem. *Let f be periodic, piecewise continuous. Then*

$$P_r * f$$

converges to f as $r \to 1$, uniformly on every compact set where f is continuous.

Next we shall see that the Poisson example involves an additional structure, namely a partial differential equation. We start by making some general comments about differentiation of the convolution integral

$$(g * f)(x) = \int g(x - t) f(t) \, dt,$$

where g is an infinitely differentiable function. We can work in either case, on the real line $\mathbf{R}$, or on the interval $[-\pi, \pi]$ if dealing with periodic functions. On the real line, we must assume that the integrals to be considered are absolutely convergent. Let D denote the derivative, which with the above choice of variables is $D = d/dx$. Then under suitable absolute convergence conditions, we can differentiate under the integral sign, and we obtain the formula

$$D(g * f) = (Dg) * f,$$

or in terms of x,

$$\frac{d}{dx} \int g(x - t) f(t) \, dt = \int \frac{d}{dx} g(x - t) f(t) \, dt.$$

Iterating the derivative gives for every positive integer m,

$$D^m(g * f) = (D^m g) * f,$$

under suitable absolute convergence conditions.

We apply all this to the function $P(r, \theta)$. Let Δ be the **Laplace operator**, which in rectangular coordinates is

$$\Delta = \left(\frac{\partial}{\partial x}\right)^2 + \left(\frac{\partial}{\partial y}\right)^2.$$

Actually, we need the expression for the Laplace operator in polar coordinates. Who knows the expression?

No answer.

I always ask this question in my classes, and nobody knows the answer. It's no different here than anywhere else. Actually, I didn't know the answer either until I had to teach it in calculus courses. Anyhow, the answer is

$$\Delta = \left(\frac{\partial}{\partial r}\right)^2 + \frac{1}{r}\left(\frac{\partial}{\partial r}\right) + \frac{1}{r^2}\left(\frac{\partial}{\partial \theta}\right)^2.$$

It's an exercise to derive this, using $x = r\cos\theta$, $y = r\sin\theta$, and the chain rule. Then you can differentiate the series for $P(r, \theta)$ term by term, with $\partial/\partial r$ or $\partial/\partial\theta$, and iterates of these partial derivatives, and you will find

$$(\Delta P)(r, \theta) = 0.$$

That the term by term differentiation is legitimate comes from the fact that the series and its partial derivatives is uniformly absolutely convergent for r with $0 \leq r \leq r_0 < 1$, and all θ.

An infinitely differentiable function f satisfying $\Delta f = 0$ is called **harmonic**. For the integral over $[-\pi, \pi]$ expressing the convolution $P_r * f$, we can differentiate under the integral sign, and so

$$\Delta(P * f)(r, \theta) = \Delta \int_{-\pi}^{\pi} P(r, \theta - t) f(t) \, dt$$

$$= \int_{-\pi}^{\pi} (\Delta P)(r, \theta - t) f(t) \, dt$$

$$= 0.$$

Hence $P * f$ is harmonic.

The above computations have applications throughout mathematics and other subjects. The physicists are particularly interested in harmonic functions, and in what is called solving the boundary value problem on the circle. Suppose you are given a continuous function f on the circle, which is the boundary of the disc of radius 1, centered at the origin. Thus f can be viewed as a function $f(\theta)$ of the variable θ, interpreted as the usual angle.

Physicists want to find a harmonic function F on the whole disc D, so a function $F(r, \theta)$ with $0 \leq r < 1$, which extends to a continuous function of the closed disc and has the prescribed value $f(\theta)$ on the boundary, so

$$\lim_{r \to 1} F(r, \theta) = f(\theta)$$

for each θ. We can now solve this problem. We let

$$F(r, \theta) = (P_r * f)(\theta) = (P * f)(r, \theta) \qquad \text{for } \theta \in \mathbf{R}.$$

By **DIR 3**, we know that

$$\lim_{r \to 1} (P * f)(r, \theta) = f(\theta),$$

so F has the continuous boundary value $f(\theta)$ for each θ. Furthermore, we have seen that $\Delta F = 0$ in the interior of the disc, so F is harmonic, as desired.

Harmonic functions on the upper half plane

The upper half plane consists of all points $(x, y) \in \mathbf{R}^2$ with x arbitrary and $y > 0$. We shall exhibit a Dirac family suited to the upper half plane, whose boundary consists of the real line, with $y = 0$. Thus we define a **Dirac family** $\{K_y\}$ on $\mathbf{R}$ with $y > 0$ as before, except that $y \to 0$ instead of $r \to 1$. Here is an example. We let

$$K_y(x) = \frac{1}{\pi} \frac{y}{x^2 + y^2} \qquad \text{with } y \to 0.$$

All the functions K_y are > 0, and a simple change of variables in the integral shows that **DIR 2** is satisfied. Also a simple estimate shows that **DIR 3** is satisfied. We leave the verification to you.

The general approximation theorem for Dirac families tells us that for every continuous bounded function f on $\mathbf{R}$, the convolution

$$(K_y * f)(x)$$

converges to $f(x)$, and does so uniformly for x in a compact set.

We have the same additional structure as for the Poisson kernel. Here we use the Laplace operator in rectangular coordinates

$$\Delta = \left(\frac{\partial}{\partial x} \right)^2 + \left(\frac{\partial}{\partial y} \right)^2.$$

Simple differentiation shows that the function $(x, y) \mapsto K_y(x)$ satisfies the Laplace equation, that is

$$\Delta K = 0.$$

In other words, K is a harmonic function on the upper half plane. By differentiating under the integral sign just as in the preceding example, we conclude that the convolution $K * f$ is harmonic for every bounded continuous function f. So we have proved the following result:

Theorem. *Let f be a bounded continuous function on the real axis. Define*

$$F(x, y) = (K_y * f)(x) = \frac{1}{\pi} \int_{\mathbf{R}} \frac{y}{(x - t)^2 + y^2} f(t) \, dt.$$

Then F is harmonic on the upper half plane, that is $\Delta F = 0$ on the upper half plane, and

$$\lim_{y \to 0} F(x, y) = f(x) \qquad \text{for all } x \in \mathbf{R}.$$

The theorem says that the given function f is the boundary value of the harmonic function F on the upper half plane, so we have constructed explicitly a harmonic function with a given boundary value.

It's nice to have given an independent construction in the present case, but if you have had a course in complex analysis, you should know that the disc and upper half plane are analytically isomorphic. Indeed, the map

$$z \mapsto \frac{z - i}{z + i}$$

gives an analytic isomorphism of the upper half plane with the unit disc. Then the Poisson kernel of the previous example corresponds to the kernel $K(x, y)$ under this isomorphism. It's a good exercise to verify this assertion in connection with a course in complex analysis.

The Heat Kernel on the Real Line

We are going to give one of the most important examples of Dirac families, if not the most important example. We now call the variables t, x and $t > 0, t \mapsto 0$. Thus a Dirac family is a family of continuous functions $\{K_t\}$ on $\mathbf{R}$, indexed by the real positive numbers t, and satisfying the following properties:

DIR 1. We have $K_t \geq 0$ for all $t > 0$.

DIR 2. $\int_{\mathbf{R}} K_t(x)\,dx = 1$ for all $t > 0$.

DIR 3. Given ϵ and δ there exists $t_0 > 0$ such that if $t < t_0$, then

$$\int_{|x| \geq \delta} K_t(x)\,dx < \epsilon.$$

We also write $K_t(x) = K(t, x)$. The general approximation theorem for a Dirac family $\{K_t\}$ is valid, and asserts:

Let f be a bounded function on $\mathbf{R}$, piecewise continuous on every finite interval. Then $(K_t * f)(x)$ converges uniformly to $f(x)$ as $t \to 0$, on every compact subset where f is continuous, or on every subset where f is uniformly continuous.

Theorem. *For $t > 0$, let*

$$K_t(x) = K(t, x) = \frac{1}{(4\pi t)^{1/2}} e^{-x^2/4t}.$$

Then $\{K_t\}$ is a Dirac family.

The verification of the axioms presents no problem. For **DIR 1** it is clear that $K(t, x) > 0$ for all t, x. For **DIR 2**, we have to prove that

$$\int_{-\infty}^{\infty} K(t, x)\,dx = \int_{-\infty}^{\infty} \frac{1}{(4\pi t)^{1/2}} e^{-x^2/4t}\,dx = 1.$$

We make a change of variables, $y^2 = x^2/4t$ so $y = x/2t^{1/2}$, $dy = dx/2t^{1/2}$. Then we have to use a standard value from integral calculus, namely

$$\int_{-\infty}^{\infty} e^{-y^2}\,dy = \sqrt{\pi},$$

and **DIR 2** falls out.

DIR 3 is not hard either, and we have to estimate a little. Given ϵ, δ we have to show that for t sufficiently close to 0,

$$(*) \qquad \frac{1}{2t^{1/2}} \int_{\delta}^{\infty} e^{-x^2/4t}\,dx < \epsilon.$$

We change variables, putting $x = 2t^{1/2}y$ as above. Then the integral $(*)$ to be estimated becomes

$$\int_{\delta/2t^{1/2}}^{\infty} e^{-y^2} dy.$$

As $t \to 0$, the lower bound of integration $\delta/2t^{1/2}$ tends to ∞, and since

$$\int_{-\infty}^{\infty} e^{-y^2} dy \quad \text{is finite,}$$

It follows that

$$\lim_{A \to \infty} \int_{A}^{\infty} e^{-y^2} dy = 0.$$

This proves **DIR 3**.

The function $(t, x) \mapsto K(t, x)$ is called the **heat kernel** on **R**. The terminology comes from physics, and may suggest that it has something to do with heat, which it may, but from our point of view, $K(t, x)$ is just a major example of Dirac families which comes up all over the place, whether it has to do with heat or not.

There is also a structure of a differential equation. We define the differential operator **H** on $\mathbf{R}^+ \times \mathbf{R}$ to be

$$\mathbf{H}_{t,x} = -\left(\frac{\partial}{\partial x}\right)^2 + \frac{\partial}{\partial t}.$$

We call **H** the **heat operator**.

Theorem. *The heat kernel satisfies the heat equation, that is*

$$\mathbf{H}K = 0, \quad \text{or in terms of the variables} \quad \mathbf{H}_{t,x} K(t, x) = 0.$$

To see this, keep cool, calm, and collected. Differentiate with respect to x twice and with respect to t once. You will get the same value, unless you make a computational error, which I often do. When you subtract, you get 0.

Differentiating under the integral sign yields:

Corollary. *Let f be a bounded continuous function on **R**. Let*

$$F(t, x) = (K_t * f)(x).$$

Then $\mathbf{H}F = 0$, i.e. F satisfies the heat equation.

This is just a special case of the general possibility of differentiating a convolution under the integral sign, that is

$$D(g * f) = (Dg) * f$$

under conditions of absolute convergence. These conditions are satisfied here.

Thus we see that we have solved a boundary value problem on the upper half plane, with given boundary value f, for the heat operator rather than the Laplace operator which we had encountered before. One says that K is the **fundamental solution** of the heat equation, because convolution with K provides such a solution with given boundary values, by convolution.

It is not more difficult to work on $\mathbf{R}^n$ rather than $\mathbf{R}$. One uses the n-dimensional version

$$K_t^{\mathbf{R}^n}(x) = K^{\mathbf{R}^n}(t, x) = \frac{1}{(4\pi t)^{n/2}} e^{-x^2/4t},$$

where $x = (x_1, \ldots, x_n)$ is an n-tuple of real numbers, and

$$x^2 = x \cdot x = x_1^2 + \cdots + x_n^2$$

is the usual dot product of x with itself. The verification that the three conditions **DIR 1, DIR 2, DIR 3** are satisfied is routine, just as in the one-dimensional case. Don't get entangled taking partial derivatives.

The heat operator on $\mathbf{R}^n$ is then

$$\mathbf{H}_{t,x} = -\Delta_x + \frac{\partial}{\partial t},$$

where

$$\Delta_x = \left(\frac{\partial}{\partial x_1}\right)^2 + \cdots + \left(\frac{\partial}{\partial x_n}\right)^2$$

is the Laplace operator. These can be applied to a function

$$F(t, x) = F(t, x_1, \ldots, x_n).$$

The Heat Kernel on the Circle

Just as we considered periodic functions when we discussed Fourier series, we consider periodic functions in connection with the heat kernel. So we want a Dirac family for periodic functions, satisfying the heat equation. The heat operator is the same as on **R**. We want a family $\{K_t\}$ of periodic functions

$$K(t, x) = K_t(x) \qquad \text{satisfying} \qquad K(t, x + 2\pi) = K(t, x),$$

and also the following three conditions:

DIR 1. $K_t \geq 0$ for all $t > 0$.

DIR 2. $\int_0^{2\pi} K_t(x)\,dx = 1$ for all $t > 0$.

DIR 3. Given ϵ, δ between 0 and 1, there exists t_0 with $0 < t_0 < 1$ such that for $0 < t < t_0$ we have

$$\int_{-\pi}^{-\delta} K_t(x)\,dx + \int_{\delta}^{\pi} K_t(x)\,dx < \epsilon.$$

In addition, if we let

$$\mathbf{H} = -\left(\frac{\partial}{\partial x}\right)^2 + \frac{\partial}{\partial t},$$

then we also want K to satisfy the partial differential equation

$$\mathbf{H}K = 0.$$

To distinguish such a periodic function K from the non-periodic heat kernel on **R**, we shall write $K^{\mathbf{R}}$ for the heat kernel on **R**, that is

$$K^{\mathbf{R}}(x) = \frac{1}{(4\pi t)^{1/2}} e^{-x^2/4t}.$$

It is natural to construct a periodic solution to our problem by periodizing the non-periodic solution on **R**. Thus we define

$$(1) \qquad K^{\mathbf{S}}(t, x) = \sum_{-\infty}^{\infty} K^{\mathbf{R}}(t, x + 2\pi n) = \sum_{n \in \mathbf{Z}} K^{\mathbf{R}}(t, x + 2\pi n).$$

The superscript **S** denotes the circle. It is standard to denote the integers by **Z**, and the sum is taken over all integers $n \in \mathbf{Z}$.

For each value of $t > 0$, the terms in the series (1) have exponential decay, so the series converges very fast. We let you write down the details.

It then follows that $K^S(t, x)$ is periodic, that is

$$K^S(t, x + 2\pi) = K^S(t, x)$$

for all $t > 0$, and $x \in \mathbf{R}$.

Theorem. *The function K^S defines a Dirac family of periodic functions, and satisfies the heat equation $\mathbf{H}K^S = 0$.*

Proof. The positivity condition **DIR 1** is obviously satisfied, even in the usual strong form that $K^S(t, x) > 0$ for all t, x (strict positivity).

For **DIR 2**, we reduce the integral over $[0, 2\pi]$ to an integral over $\mathbf{R}$. We have

$$
\begin{aligned}
\int_0^{2\pi} K^S(t, x)\,dx &= \int_0^{2\pi} \frac{1}{(4\pi t)^{1/2}} \sum_{n \in \mathbf{Z}} e^{-(x+2\pi n)^2/4t}\,dx \\
&= \frac{1}{(4\pi t)^{1/2}} \sum_{n \in \mathbf{Z}} \int_0^{2\pi} e^{-(x+2\pi n)^2/4t}\,dx \\
&= \frac{1}{(4\pi t)^{1/2}} \sum_{n \in \mathbf{Z}} \int_{2\pi n}^{2\pi(n+1)} e^{-y^2/4t}\,dy \\
&= \frac{1}{(4\pi t)^{1/2}} \int_{-\infty}^{\infty} e^{-y^2/4t}\,dt \\
&= \int_{-\infty}^{\infty} K_t^R(y)\,dy \\
&= 1.
\end{aligned}
$$

Thus we have reduced **DIR 2** on the circle to **DIR 2** on the line, for the heat kernel K^R on the line, whose integral we already had seen to be equal to 1.

There remains to prove **DIR 3**. Again, we reduce the property to the corresponding property of K^R on the line. We have to show that the integral of K_t^S on an interval $[\delta, \pi]$ approaches 0 when t approaches 0. To do this, we again exchange a sum and an integral, namely

$$
\frac{1}{(4\pi t)^{1/2}} \sum_{n=0}^{\infty} e^{-(x+2\pi n)^2/4t}\,dx = \sum_{n=0}^{\infty} \frac{1}{(4\pi t)^{1/2}} \int_{\delta+2\pi n}^{2\pi(n+1)} e^{-y^2/4t}\,dy
$$

$$
\leqq \int_\delta^{\infty} K^R(t, y)\,dy.
$$

By **DIR 3** for the heat kernel on $\mathbf{R}$, we know that there exists t_0 with $0 < t_0 < 1$ such that this last expression for $0 < t < t_0$ is $< \epsilon$. An analogous argument proves the inequality for the interval $[-\pi, -\delta]$. Thus we have proved **DIR 3** for K^S.

Finally we look at the differential equation. The convergence of the periodized series

$$\sum_{-\infty}^{\infty} e^{-(x+2\pi n)^2/4t}$$

is sufficiently fast that we can differentiate term by term, whether we take $\partial/\partial t$ or $\partial/\partial x$ and $(\partial/\partial x)^2$. But if c is any constant, and f satisfies the heat equation, that is

$$\mathbf{H}_{t,x} f(t, x) = 0 \qquad \text{for all } t, x,$$

then the function g defined by

$$g(t, x) = f(t, x + c)$$

also satisfies the equation. Hence each term

$$f_n(t, x) = \frac{1}{(4\pi t)^{1/2}} e^{-(x+2\pi n)^2/4t}$$

satisfies the heat equation $\mathbf{H} f_n = 0$. Hence $\mathbf{H} K^S = 0$, as was to be shown.

Next we relate convolution on the circle (i.e. for periodic functions) with convolution on the real line. On the circle, convolution is given by the formula

$$f * g(x) = \int_0^{2\pi} f(y) g(x - y) \, dy,$$

for functions f, g periodic of period 2π. Let f be periodic. Then

$$K_t^S * f(x) = \int_0^{2\pi} K_t^S(y) f(x - y) \, dy$$

$$= \int_0^{2\pi} \sum_{n \in \mathbf{Z}} \frac{1}{(4\pi t)^{1/2}} e^{-(y+2\pi n)^2/4t} f(x - y) \, dy$$

$$= \frac{1}{(4\pi t)^{1/2}} \sum_{n \in \mathbf{Z}} \int_0^{2\pi} e^{-(y+2\pi n)^2/4t} f(x - y) \, dy$$

$$= \frac{1}{(4\pi t)^{1/2}} \sum_{n \in \mathbf{Z}} \int_{2\pi n}^{2\pi(n+1)} e^{-y^2/4t} f(x - y) \, dy$$

$$\text{(because } f \text{ is periodic)}$$

$$= \frac{1}{(4\pi t)^{1/2}} \int_{-\infty}^{\infty} e^{-y^2/4t} f(x - y) \, dy.$$

Thus we have related convolution on the circle with convolution on **R**. We can summarize the computation in a theorem.

Theorem. *Let f be continuous, periodic of period 2π. Then*

$$K_t^S * f(x) = K_t * f(x),$$

where the convolution on the left is on $[0, 2\pi]$, and convolution on the right is on $(-\infty, \infty)$.

Note that the formula

$$\mathbf{H}(K^S * f) = (\mathbf{H}K^S) * f = 0$$

can also be seen from this last theorem.

The Fourier series and Poisson inversion

Since K_t^S is periodic, it has a Fourier series, which can easily be computed. Actually, K_t^S is infinitely differentiable, as one verifies by differentiating term by term, and showing that the differentiated series converges absolutely and uniformly on every bounded region $|x| \leq A$. By general theorems on Fourier series, the Fourier series of K_t^S converges to the function K_t^S. The next theorem tells us what is this Fourier series.

Theorem. *The Fourier series for K_t^S is given by*

$$\frac{1}{(4\pi t)^{1/2}} \sum_{n \in \mathbf{Z}} e^{-(x+2\pi n)^2/4t} = \frac{1}{2\pi} \sum_{n \in \mathbf{Z}} e^{-n^2 t} e^{inx}.$$

Proof. The m-th Fourier coefficients is determined by the integral for $m \in \mathbf{Z}$,

$$\int_0^{2\pi} K_t^S(x) e^{-imx} dx = \sum_{n \in \mathbf{Z}} \frac{1}{(4\pi t)^{1/2}} \int_0^{2\pi} e^{-(x+2\pi n)^2/4t} e^{-imx} dx$$

$$= \sum_{n \in \mathbf{Z}} \frac{1}{(4\pi t)^{1/2}} \int_{2\pi n}^{2\pi(n+1)} e^{-y^2/4t} e^{-imy} dy,$$

where we substituted $y = x + 2\pi n$ and use $e^{im2\pi n} = 1$. The last expression is

$$= \frac{1}{(4\pi t)^{1/2}} \int_{-\infty}^{\infty} e^{-y^2/4t} e^{-imy} dy.$$

We make the substitution $y = (2t)^{1/2}u$, $dy = (2t)^{1/2}\,du$, and use a standard value from calculus, namely

$$\frac{1}{(2\pi)^{1/2}} \int_{-\infty}^{\infty} e^{-u^2/2} e^{-iuv}\,du = e^{-v^2/2}.$$

We then obtain

$$\int_0^{2\pi} K_t^S(x) e^{-imx}\,dx = e^{-m^2 t}.$$

This proves the theorem.

The theorem is due to Poisson, and is called the **Poisson inversion formula**. A very important special case comes by taking $x = 0$, so e^{inx} becomes 1. In this case, we obtain what's known as the **Poisson summation formula**.

Corollary.

$$\frac{1}{(4\pi t)^{1/2}} \sum_{n \in \mathbf{Z}} e^{-(2\pi n)^2/4t} = \frac{1}{2\pi} \sum_{n \in \mathbf{Z}} e^{-n^2 t}.$$

There are many applications for this formula in analysis and number theory. We shall give one of these applications below, showing how Riemann proved the functional equation of the zeta function. In analysis, the Poisson inversion formula comes up in spectral theory. For a very general context of inversion formulas, cf. [JoL 94] and [JoL 96].

The Fourier series of the heat kernel has a shape which is of independent interest, and is called a theta series. We shall give specific properties of such theta series in a separate section.

Theta Series and the Convolution Product

We start from the Fourier series for the heat kernel on the circle. Without knowing anything, we may define the **theta function** $\theta_t(x)$ (for $x \in \mathbf{R}$ and $t > 0$) by the series

$$\theta_t(x) = \frac{1}{2\pi} \sum_{n \in \mathbf{Z}} e^{-n^2 t} e^{inx}.$$

Of course, the periodic heat kernel K_t^S is equal to the theta series, but we don't need to know this for what comes next. We just start with the above definition, so the arguments proving the next theorem are self-contained.

Theorem. *For positive real numbers t, s we have the relation*

$$\theta_t * \theta_s = \theta_{t+s}.$$

Proof. We wrote the series for $\theta_t(x)$ above, and we now write the series with s instead of t,

$$\theta_s(x) = \frac{1}{2\pi} \sum_{m \in \mathbf{Z}} e^{-m^2 s} e^{imx}.$$

Because of the rapid decrease of $e^{-n^2 t}$ and $e^{-m^2 s}$ the convolution is given by

$$\theta_t * \theta_s(x) = \int_0^{2\pi} \theta_t(x - y)\theta_s(y)\,dy$$

$$= \frac{1}{(2\pi)^2} \sum_{m,n} \int_0^{2\pi} e^{-n^2 t} e^{-m^2 s} e^{in(x-y)} e^{imy}\,dy$$

$$= \frac{1}{(2\pi)^2} \sum_{m,n} e^{inx} e^{-(n^2 t + m^2 s)} \int_0^{2\pi} e^{i(m-n)y}\,dy.$$

The usual argument shows that the last integral is 0 if $m \neq n$, and is 2π if $m = n$. Hence only terms with $m = n$ remain in the sum, and we get

$$\theta_t * \theta_s(x) = \frac{1}{(2\pi)^2} \sum_{n \in \mathbf{Z}} e^{-n^2(t+s)} \cdot (2\pi)$$

$$= \theta_{t+s}(x),$$

which proves the theorem.

Remarks. Of course, since the theta function is equal to the heat kernel, the same formula applies to the heat kernel. In other words,

$$K_t^S * K_{t'}^S = K_{t+t'}^S.$$

The convolution operator is associative, that is for three functions h, g, f we have

$$(h * g) * f = h * (g * f).$$

You can verify this associativity by interchanging the order of integration. For each t, the function K_t^S can be viewed as an integral operator

$$f \mapsto K_t^S * f.$$

Then the above convolution formula $K_t^S * K_{t'}^S = K_{t+t'}^S$ tells you that the operator composed from K_t^S and $K_{t'}^S$ is given by convolution with $K_{t+t'}^S$. Thus one says that the heat kernel $\{K_t^S\}$ is a semigroup of operators, because it is closed under composition.

So far, we have given all the basic properties of the kernel on the real line. If you go forward in mathematics, you will realize that these properties are typical of very general situations. What we did was not accidental but was a prototype for bigger and better theories.

Next we leave heat kernel considerations, and point to seemingly completely different aspects of the theta function.

The Poisson Summation Formula and Functional Equation of the Zeta Function

One of the applications of the Poisson summation formula is to Riemann's proof of the functional equation of the zeta function, which we shall now give. First we introduce the function

$$\psi(t) = \sum_{n \in \mathbf{Z}} e^{-n^2 \pi t},$$

which up to a change of variables and a constant factor is the theta function. More precisely,

$$\psi(t) = 2\pi\theta(\pi t, 0).$$

Then the Poisson summation formula can be rewritten in the form

(1) $$\psi(t^{-1}) = t^{1/2}\psi(t).$$

Next the **zeta function** ζ is defined by the series

$$\zeta(s) = \sum_{n=1}^{\infty} \frac{1}{n^s}.$$

This series converges absolutely for s real > 1, as should be known from calculus and the integral test. You can also take s complex with $\mathrm{Re}(s) > 1$ if you know about complex numbers. Also the convergence is uniform for

$$\mathrm{Re}(s) \geq c > 1.$$

But we can stick to real s if you want. The problem is to give an expression for $\zeta(s)$ which will converge for all values of s with the only possible exception at $s = 0$ and $s = 1$. That's what Riemann did, and we follow Riemann. The argument depends on considering what is called the **Mellin transform** $\mathbf{M}g$ of a suitable function g, defined by

$$(\mathbf{M}g)(s) = \int_0^\infty g(t)t^s \frac{dt}{t}.$$

Still for $s > 1$ we introduce the function

$$F(s) = \pi^{-s/2}\Gamma\left(\frac{s}{2}\right)\zeta(s),$$

where the **gamma function** is defined by the usual integral

$$\Gamma(s) = \int_0^\infty e^{-t}t^s \frac{dt}{t}.$$

Theorem (Riemann). *The function F admits an analytic expression for all $s \neq 0, 1$, and satisfies the equation*

$$F(s) = F(1 - s).$$

Proof. Define the function

(2)
$$g(t) = \sum_{n=1}^\infty e^{-n^2\pi t} = \tfrac{1}{2}(\psi(t) - 1),$$

or also

$$2g(t) = \psi(t) - 1.$$

Then

$$(\mathbf{M}g)\left(\frac{s}{2}\right) = \int_0^\infty g(t)t^{s/2}\frac{dt}{t}.$$

Lemma. *We have*

$$(\mathbf{M}g)\left(\frac{s}{2}\right) = F(s) \qquad \text{for} \quad s > 1.$$

Proof. We substitute the series for $g(t)$ and interchange the order of summation and integration. We obtain

$$(\mathbf{M}g)\left(\frac{s}{2}\right) = \sum_{n=1}^\infty \int_0^\infty e^{-n^2\pi t}t^{s/2}\frac{dt}{t}.$$

Make the change of variables $t \mapsto t/n^2\pi$, that is $u = n^2\pi t$, $du = n^2\pi \, dt$. Then

$$(\mathbf{M}g)\left(\frac{s}{2}\right) = \sum_{n=1}^{\infty} \frac{1}{n^s \pi^{s/2}} \int_0^{\infty} e^{-u} u^{s/2} \frac{du}{u}$$

$$= \sum_{n=1}^{\infty} \pi^{-s/2} \frac{1}{n^s} \Gamma\left(\frac{s}{2}\right)$$

$$= F(s),$$

which proves the lemma.

Next, we decompose the Mellin integral $(\mathbf{M}g)(s/2)$ into two integrals,

$$(\mathbf{M}g)\left(\frac{s}{2}\right) = \int_1^{\infty} g(t) t^{s/2} \frac{dt}{t} + \int_0^1 g(t) t^{s/2} \frac{dt}{t}$$

$$= \int_1^{\infty} g(t) t^{s/2} \frac{dt}{t} + \int_1^{\infty} g\left(\frac{1}{t}\right) t^{-s/2} \frac{dt}{t},$$

after making the transformation $t \mapsto 1/t$. By (2) we obtain

$$g\left(\frac{1}{t}\right) = \frac{1}{2}\left(\frac{1}{t} - 1\right) = \frac{1}{2} t^{1/2} \psi(t) - \frac{1}{2} = t^{1/2} g(t) + \frac{1}{2} t^{1/2} - \frac{1}{2}.$$

Substituting this value in the integral containing $g(1/t)$, and taking into account that $F(s) = (\mathbf{M}g)(s/2)$, we get

$$F(s) = \frac{1}{s-1} - \frac{1}{s} + \int_1^{\infty} g(t)\left(t^{s/2} + t^{(1-s)/2}\right) \frac{dt}{t}.$$

The first two terms come from direct integration and calculus. Now we observe that the integral expression is valid for all values of s, and the first two terms exhibit singularities at $s = 1$ and $s = 0$, but otherwise are nice simple functions. Furthermore, interchanging s and $1 - s$ leaves the expression

$$\frac{1}{s-1} - \frac{1}{s}$$

unchanged, and also leaves the expression under the integral sign unchanged. This concludes the proof of the functional equation.

Postscript. Incidentally, if you now go back to the counting of prime numbers, and the Riemann hypothesis, you will see the same zeta function that is occurring here. By pushing the sort of analysis somewhat further, Riemann obtained an explicit formula for $\pi(x)$ in terms of the zeros of this zeta function. The equivalence between the Riemann hypothesis as we stated it in terms of counting, and the way Riemann stated it, is a direct consequence of this explicit formula. But carrying out the explicit formula is beyond the level of the present talks.

Theta Functions and Complex Doubly Periodic Functions

Just to show how branches of mathematics which at first may appear different actually are related, and it's all the same thing, I shall now describe the way theta functions appear to factorize certain types of complex analytic functions, called elliptic functions. This factorization is analogous to the factorization of a polynomial into linear factors over the complex numbers. At a deeper level, it is analogous to the factorization of a positive integer into primes. But to expand on these analogies takes more and more space and time, and we have to stop somewhere, so we stop after the present discussion.

Everybody knows the periodic functions sine and cosine, which are related to the geometry of the circle. Beyond this geometry, you also know from calculus that they satisfy the differential equation

$$(1) \qquad y'^2 = 1 - y^2,$$

which is just another way of writing

$$\sin'(x) = \cos x \qquad \text{and} \qquad \sin^2(x) + \cos^2(x) = 1.$$

Furthermore, $\sin x$ and $\cos x$ are periodic, with one basic period 2π. In the eighteenth and nineteenth centuries, mathematicians came across a similar phenomenon but somewhat more complicated, namely functions f which satisfy the differential equation

$$(2) \qquad f'^2 = f^3 + Af + B.$$

Here, A and B are constants. In the differential equation for sine and cosine, the polynomial on the right side has degree 2. In the similar but more complicated situation of (2), the polynomial has degree 3. The polynomial differential equation is actually quite general, because by a change of variables, a polynomial of degree 3 can always be turned into another one which does not have the square term, i.e. of the form

$$P(T) = T^3 + AT + B,$$

where the coefficient of T^2 is 0. So the equation (2) is a direct generalization of (1) to a degree 3 polynomial on the right side. One was led to such a differential equation in the theory of complex doubly periodic functions, i.e. functions $f(z)$ of a complex variable which have two periods, ω_1, ω_2 linearly independent over $\mathbf{R}$. For instance, $\omega_1 = 1$ and $\omega_2 = i$ would be such a pair of periods. One wants to study analytic functions f such that

$$f(z + \omega_1) = f(z) \qquad \text{and} \qquad f(z + \omega_2) = f(z)$$

for all complex z. Actually, such functions must have poles if they are not constant, so we are really dealing with what's called meromorphic functions having two independent periods. Such functions are called **elliptic**. It's remarkable that the study of elliptic functions leads to theta functions as we shall describe below. These theta functions for certain values of the variables satisfy the heat equation, and as we have seen before, are the fundamental solution of the heat equation on the circle.

For applications to the complex situation, one defines from scratch the **Riemann theta function**

$$\theta_1(\tau, z) = \sum_{n \in \mathbf{Z}} e^{\pi i n^2 \tau} e^{2\pi i n z}.$$

In this definition, τ, z are complex numbers. We write $\tau = u + it$, with u, t real, so $t = \mathrm{Im}(\tau)$ is the imaginary part of τ. We assume $t > 0$, in order to make the series convergent. With this assumption, we see that

$$e^{\pi i n^2 \tau} = e^{-\pi n^2 t} e^{\pi i n^2 u}.$$

Since $e^{\pi i n^2 u}$ has absolute value 1, the absolute value of $e^{\pi i n^2 \tau}$ is just $e^{-\pi n^2 t}$. Hence the series for $\theta_1(\tau, z)$ converges absolutely, and uniformly for $|t| \geq \delta > 0$. For $z = x$ real, and $\tau = it$ pure imaginary, we essentially get the theta series which we considered previously in connection with the heat kernel on the circle.

The Riemann theta function θ_1 has a simple nearly periodic relation which we give explicitly before pointing out the application to elliptic functions. First, note that 1 is a genuine period, that is

PER 1. $\theta_1(\tau, z + 1) = \theta_1(\tau, z).$

This is immediate because

$$\theta_1(\tau, z + 1) = \sum_{n \in \mathbf{Z}} e^{\pi i n^2} \cdot e^{2\pi i n(z+1)}$$

and

$$e^{2\pi i n(z+1)} = e^{2\pi i n z} \cdot e^{2\pi i n} = e^{2\pi i n z},$$

because $e^{2\pi i n} = 1$ for all $n \in \mathbf{Z}$. Next, τ is not quite a period, but still, we have:

PER 2. $\qquad \theta_1(\tau_1, z + \tau) = e^{-\pi i(\tau + 2z)} \cdot \theta_1(\tau, z).$

The computation is short, as follows.

$$
\begin{aligned}
\theta_1(\tau, z + \tau) &= \sum_{n \in \mathbf{Z}} e^{\pi i n^2 \tau} \cdot e^{2\pi i n z} \cdot e^{2\pi i n \tau} \\
&= \sum_{n \in \mathbf{Z}} e^{\pi i (n^2 + 2n + 1)\tau} \cdot e^{-\pi i \tau} \cdot e^{2\pi i n z} \\
&= e^{-\pi i(\tau + 2z)} \sum_{n \in \mathbf{Z}} e^{\pi i (n+1)^2 \tau} \cdot e^{2\pi i (n+1)z} \\
&= e^{-\pi i(\tau + 2z)} \cdot \theta_1(\tau, z),
\end{aligned}
$$

thus proving **PER 2**.

The two relations **PER 1** and **PER 2** allow us to construct doubly periodic functions. All we have to do is concoct something which gets rid of the extra factor coming in **PER 2**. Let $a_1, \ldots, a_n$ and $b_1, \ldots, b_n$ be complex numbers with the property that

$$a_1 + \cdots + a_n = b_1 + \cdots + b_n.$$

Let

$$f(z) = \frac{\prod_j \theta_1(\tau, z - a_j)}{\prod_j \theta_1(\tau, z - b_j)}.$$

Then you see from **PER 1** that $f(z + 1) = f(z)$, and because of our assumption on $a_1, \ldots, a_n$, $b_1, \ldots, b_n$ it also drops out of **PER 2** that $f(z + \tau) = f(z)$. Thus we have constructed lots of doubly periodic functions, with periods $1, \tau$.

One can show that all doubly periodic meromorphic functions can be constructed in the above manner with theta functions slightly more generally than we have given above, but by an entirely similar procedure. For this, look up books on elliptic functions, e.g. [La 87].

To see other connections of theta functions, look up the survey of Gesztesy and Weikard [GeW 98], which came out when the present book was in production.

Bibliography

[GeW 98]   F. GESZTESY and R. WEIKARD, Elliptic algebro geometric solutions of the KDV and AKNS hierarchies – an analytic approach, *Bull. AMS* **35** No. 4 (1998) pp. 271–317

[JoL 94]   J. JORGENSON and S. LANG, *Explicit formulas for regularized series and products*, Springer Lecture Notes 1593, 1994

[La 83/97] S. LANG, *Undergraduate Analysis*, Springer-Verlag, 1983; Second edition, 1997

[La 87]   S. LANG, *Elliptic Functions*, Springer-Verlag, 1987

[La 99]   S. LANG, *Complex Analysis*, Fourth edition, Springer-Verlag, 1999

Bruhat–Tits Spaces

The notions and auxiliary theorems which we shall use are all standard from basic undergraduate courses in linear algebra and analysis. However, such courses are structured to provide broad foundations, and there is no time for them to go deeper into certain directions. The present exposition provides an opportunity to exhibit such a direction for students at the undergraduate level by putting together notions which are separated in the courses, such as the derivative as linear map, the exponential series applied to linear maps, positive definite matrices, length of curves, etc. One way to use the present material is to ask interested students themselves to read it and present it to others in a special undergraduate colloquium, seminar, or math club. I have done so successfully at Yale.

The first part on the semi parallelogram law can be the topic of one talk. It can be covered at the beginning of an undergraduate analysis course, just after the definition of a complete metric space. For students who know what a group is, or want to learn right away, I recommend the Bruhat–Tits fixed point theorem. There may even be something marketable here. It's well known that some fixed point theorems are used frequently not only in mathematics but also in economics. There should be some application of the Bruhat–Tits fixed point theorem in economics. It is very likely that if you find one, you will get a Nobel prize.

The second and third parts require somewhat more knowledge from an analysis course and linear algebra. It puts into practice something we tell students, that the derivative of a differentiable map at a point is a linear map, but most mathematicians follow this up by still taking partial derivatives instead of using the linear map as such. The second part defines a metric on the space of real positive definite matrices. The third part proves that this space satisfies the semi parallelogram law. Both these parts give practice to the calculus in vector spaces, mixed with some linear algebra involving scalar products. There is no time or opportunity to cover this material in standard courses, but the material gives the most classical example of some aspects of analysis and algebra, and also of geometry by showing how the exponential map "bends" ordinary euclidean space and affects distances—actually increases distances. Thus the second and third part provide examples of differential geometry, quite different from the usual examples emphasizing curves and surfaces. It shows the calculus in vector spaces at work. These parts provide a beautiful combination of analysis, linear algebra, calculus in vector spaces, and geometry at the undergraduate level. All three parts give an opportunity for a series of talks.

At the end, I shall describe some of the history behind the mathematics we are considering here. It turns out that completely elementary, beautiful, and powerful results could be extracted from a rather fancy context, as I have done. Even the history of this mathematical context is not elementary. That's life, but one should not let the deep history hide or obstruct the mathematics, accessible at an elementary level.

§1. The Semi Parallelogram Law

The ordinary parallelogram law is first encountered in the context of the euclidean plane, but is valid in a vector space over the reals with a positive definite scalar product. From such a product, we get a norm on the vector space in the usual way, namely for a vector v, the norm is defined by

$$|v| = \sqrt{v \cdot v}.$$

The parallelogram law states that for two vectors v, w one has

$$|v - w|^2 + |v + w|^2 = 2|v|^2 + 2|w|^2.$$

In words, the sums of the squares of the diagonals is equal to twice the sum of the squares of the sides. The parallelogram law can be illustrated as in the figure. It is well known that a norm comes from a positive definite scalar

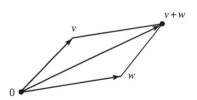

product if and only if the norm satisfies the parallelogram law, but we are not concerned here with this direction. We shall develop some mathematics in reverse order to the historical order, and then make appropriate historical comments.

In the formula given above for the parallelogram law, the parallelogram is located at the origin. The law actually relates the lengths of the sides with the lengths of the diagonals. It applies to a parallelogram located at any point x, where the picture is as follows:

The distance between two points x_1, x_2 is given by

$$d(x_1, x_2) = |x_1 - x_2|.$$

Then the parallelogram law states that

$$d(x_1, x_2)^2 + d(x, x_3)^2 = 2d(x, x_1)^2 + 2d(x, x_2)^2.$$

The parallelogram law will be semified in two ways: first, we stop at the midpoint of one of the diagonals; and second, we write an inequality instead of the equality. As to the first, we draw half the parallelogram:

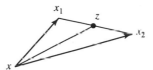

Let z be the midpoint between x_1 and x_2. Then in terms of x, x_1, x_2, and z, the parallelogram law can be written down in the form

$$d(x_1, x_2)^2 + 4d(x, z)^2 = 2d(x, x_1)^2 + 2d(x, x_2)^2.$$

Next we semify by writing down an inequality $\leq$ instead of the equality, and we also go over to more general spaces, so let's start from scratch.

Let X be a metric space, with a distance function d, so the distance $d(x, y)$ is defined between two points x, y; it is ≥ 0 and $= 0$ if and only if $x = y$; we have

$$d(x, y) - d(y, x),$$

and the distance satisfies the triangle inequality

$$d(x, z) \leq d(x, y) + d(y, z)$$

for all points x, y, $z \in X$. A priori, a midpoint between two points x_1, x_2 is not defined. We shall now describe a condition which will give rise to a midpoint.

We say that X satisfies the **semi parallelogram law** if for any two points x_1, $x_2 \in X$ there is a point z such that for all $x \in X$,

$$d(x_1, x_2)^2 + 4d(x, z)^2 \leq 2d(x, x_1)^2 + 2d(x, x_2)^2.$$

It follows that

$$d(z, x_1) = d(z, x_2) = \tfrac{1}{2}d(x_1, x_2).$$

This is obtained by setting $x = x_1$ and $x = x_2$ in the semi parallelogram law to get the inequalities $2d(x_1, z) \leqq d(x_1, x_2)$ and $2d(x_2, z) \leqq d(x_1, x_2)$. The opposite inequalities follow from the triangle inequality

$$d(x_1, x_2) \leqq d(x_1, z) + d(z, x_2).$$

Note that the point z is uniquely determined by x_1, x_2 because if z' is another such point, we put $x = z'$ in the law to see that $z = z'$. Thus we call z the **midpoint** between x_1 and x_2.

A **Bruhat–Tits space** is defined to be a complete metric space which satisfies the semi parallelogram law. "Complete" means that every Cauchy sequence in X converges, i.e. has a limit in X. We denote by $\mathbf{B}_R(x)$ the open ball of radius R centered at x, and by $\bar{\mathbf{B}}_R(x)$ the close ball of radius R centered at x. So $\bar{\mathbf{B}}_R(x)$ is the set of all points $y \in X$ such that

$$d(x, y) \leqq R.$$

Let X be a metric space. Let S be a bounded subset, i.e. there exists a closed ball $\bar{\mathbf{B}}_R(x)$ for some x such that $S \in \bar{\mathbf{B}}_R(x)$. Let

$r = $ inf of all R such that $S \subset \bar{\mathbf{B}}_R(x)$ for all possible radii R and centers $x \in X$.

It may very well be that there is no closed ball of radius r containing S. If such a ball of radius r exists, then we call it a ball of minimal radius containing S. We are now going to give a condition that such a ball exists.

Theorem 1.1 (Serre). *Let X be a Bruhat–Tits space. Let S be a bounded subset of X. Then there exists a unique closed ball $\bar{\mathbf{B}}_r(x_1)$ of minimal radius containing S.*

Proof. We first prove uniqueness. Suppose there are two balls $\bar{\mathbf{B}}_r(x_1)$ and $\bar{\mathbf{B}}_r(x_2)$ of minimal radius containing S, but $x_2 \neq x_1$. Let x be any point of S, so $d(x, x_2) \leqq r$ and $d(x, x_1) \leqq r$. Let z be the midpoint between x_1 and x_2. By the semi parallelogram law, we have

$$d(x_1, x_2)^2 \leqq 4r^2 - 4d(x, z)^2.$$

Since by definition of r, there are points $x \in S$ such that $d(x, z) \geqq r - \epsilon$

for all $\epsilon > 0$, it follows that

$$d(x_1, x_2)^2 \leqq 4r^2 - 4(r - \epsilon)^2$$

for every $\epsilon > 0$. Taking the limit as $\epsilon \to 0$, we conclude that $d(x_1, x_2)^2 = 0$, whence $x_1 = x_2$, so the centers of the balls are equal. Since the balls have the same radius r, the balls are equal, thus proving the uniqueness.

As to existence, let $\{x_n\}$ be a sequence of points which are centers of balls of radius r_n approaching the inf of all such radii such that $\bar{\mathbf{B}}_{r_n}(x_n)$ contains S. Let r be this inf. If the sequence $\{x_n\}$ is a Cauchy sequence, then it converges to some point which is the center of a closed ball of the minimal radius containing S, and we are done. It's easy to put in the details of this argument, and we leave these details to you.

The main point is to show that the centers do form a Cauchy sequence, i.e. we are always in the above situation. Even in the plane, there is something to prove which is relevant to high school plane geometry. We give the proof.

Let z_{mn} be the midpoint between x_n and x_m. By the minimality of r, given ϵ there exists a point $x \in S$ such that

$$d(x, z_{mn})^2 \geqq r^2 - \epsilon.$$

We apply the semi parallelogram law with $z = z_{mn}$. Then

$$d(x_m, x_n)^2 \leqq 2d(x, x_m)^2 + 2d(x, x_n)^2 - 4d(x, z_{mn})^2$$
$$\leqq 2r_m^2 + 2r_n^2 - 4r^2 + 4\epsilon.$$

As $m, n \to \infty$, $2r_m^2 + 2r_n^2 - 4r^2 \to 0$. Then the above inequality shows that $\{x_n\}$ is Cauchy, thus concluding the proof of the theorem.

The center of the ball in Theorem 1.1 is called the **circumcenter** of the set S.

Let X be a metric space. By an **isometry** of X we mean a bijection

$$g: X \to X$$

such that g preserves distances. In other words, for all $x_1, x_2 \in X$ we have

$$d(g(x_1), g(x_2)) = d(x_1, x_2).$$

If plane geometry was properly taught in high school, you should know that translations, rotations, and reflections are isometries of the euclidean plane, and that all isometries of the plane can be obtained by composition of these special ones. In any case, these mappings provide examples of isometries. Note that if g_1, g_2 are isometries, so is the composite $g_1 \circ g_2$. Also if g is an isometry, then g has an inverse mapping (because g is a bijection), and the isometry condition immediately shows that $g^{-1} \colon X \to X$ is also an isometry. Note that the identity mapping $\mathrm{id} \colon X \to X$ is an isometry.

Let G be a set of isometries. We say that G is a **group** of isometries if G contains the identity mapping, G is closed under composition (that is, if $g_1, g_2 \in G$ then $g_1 \circ g_2 \in G$), and is closed under inverse (that is, if $g \in G$ then $g^{-1} \in G$). One often writes $g_1 g_2$ instead of $g_1 \circ g_2$. Note that the set of all isometries is itself a group of isometries.

Let $x' \in X$. The subset Gx' consisting of all elements $g(x')$ with $g \in G$ is called the **orbit** of x' under G. Let S denote this orbit. Then for all $g \in G$ and all elements $x \in S$ it follows that $gx \in S$. Indeed, we can write $x = g_1 x'$ for some $g_1 \in G$. Then

$$g(g_1 x') = g(g_1(x')) = (g \circ g_1)(x') \in S$$

and $g \circ g_1 \in G$ by assumption. In fact $g(S) = S$ because a given $g_1 x'$ can be written

$$g_1 x' = g g^{-1} g_1 x' = g x, \quad x \in S.$$

After these preliminaries, we have the following theorem:

Theorem 1.2 (Bruhat–Tits fixed point theorem). *Let X be a Bruhat–Tits metric space. Let G be a group of isometries of X with the action of G denoted by $(g, x) \mapsto g \cdot x$. Suppose G has a bounded orbit (this occurs if, for instance, G is compact and the action is continuous). Then G has a fixed point, for instance the circumcenter of the orbit.*

Proof. Let $p \in X$ and let $G \cdot p$ be the orbit. Let $\bar{\mathbf{B}}_r(x_1)$ be the unique closed ball of minimal radius containing this orbit. For any $g \in G$, the image $g \cdot \bar{\mathbf{B}}_r(x_1) = \bar{\mathbf{B}}_r(x_2)$ is a closed ball of the same radius containing the orbit, and $x_2 = g \cdot x_1$, so by the uniqueness of Theorem 1.1, it follows that x_1 is a fixed point, thus concluding the proof.

For the next corollary, we assume you know the definition of a group and the coset space G/H with respect to a subgroup H. If you don't, skip the corollary.

Corollary 1.3. *Let G be a group, H a subgroup. Let K be a subgroup of G, so that K acts by translation on the coset space G/H. Suppose G/H has a metric (distance function) such that translation by elements of K are isometries, G/H is a Bruhat–Tits space, and one orbit is bounded. Then a conjugate of K is contained in H.*

Proof. By Corollary 1.2, the action of K has a fixed point, i.e. there exists a coset xH such that $kxH = xH$ for all $k \in K$. Then $x^{-1}KxH \subset H$, whence $x^{-1}Kx \subset H$, as was to be shown.

Note that Theorems 1.1 and 1.2 are completely elementary, and could be done immediately after you know the definition of a complete metric space. What comes next is somewhat more involved, and requires a bit more linear algebra, mixed with some undergraduate analysis. So what comes next may be viewed as an interesting topic to accompany a course in advanced calculus–analysis.

§2. The Space of Positive Definite Matrices

We shall now describe the setting for the most important example of a Bruhat–Tits space. It is actually a typical example. Let:

$\text{Mat}_n(\mathbf{R}) = $ vector space of real $n \times n$ matrices.

$\text{GL}_n(\mathbf{R}) = $ set of real invertible $n \times n$ matrices.

Note that $\text{GL}_n(\mathbf{R})$ is contained in $\text{Mat}_n(\mathbf{R})$, and $\text{GL}_n(\mathbf{R})$ is closed under multiplication, and under taking the multiplicative inverse. We then say that $\text{GL}_n(\mathbf{R})$ is a group, which will be abbreviated by G, so $G = \text{GL}_n(\mathbf{R})$ by definition.

The vector space $\text{Mat}_n(\mathbf{R})$ has a subspace:

$\text{Sym}_n(\mathbf{R}) = $ space of symmetric matrices.

Recall that if v is a matrix, then its **transpose** $^t v$ is defined by interchanging rows and columns, and v is called **symmetric** when $v = {}^t v$. We often omit $\mathbf{R}$ from the notation, and write simply Mat_n, GL_n, and Sym_n. The finite dimensional vector space Sym_n has a positive definite scalar product which allows us to view Sym_n as a euclidean space. The product is defined as follows. Recall that the trace of a matrix is the sum of the diagonal elements. For $v, w \in \text{Sym}_n$, we define the **trace scalar product** to be

$$\langle v, w \rangle_{\text{tr}} = \text{tr}(vw) \qquad \text{so that} \qquad |v|_{\text{tr}}^2 = \text{tr}(v^2).$$

Here and elsewhere, tr denotes the trace of a matrix. Since v is symmetric, the trace $\text{tr}(v^2)$ is a sum of squares—check it by writing down the matrix multiplication. So it is immediately verified that $(v, w) \mapsto \text{tr}(vw)$ is a positive definite scalar product which makes Sym_n into a euclidean space of dimension $n(n + 1)/2$.

Next we recall that a matrix p is called **positive**, or **positive definite**, if it is symmetric and

$$\langle p\xi, \xi \rangle > 0 \qquad \text{for all } \xi \in \mathbf{R}^n, \, \xi \neq 0.$$

Here, $\langle \ , \ \rangle$ denotes the ordinary scalar product on $\mathbf{R}^n$, i.e. the dot product. We let:

$\text{Pos}_n = \text{Pos}_n(\mathbf{R}) = $ set of all symmetric positive definite $n \times n$ matrices.

Thus Pos_n is a subset of Sym_n. It follows immediately from the definition that Pos_n is an open subset of Sym_n, so we think geometrically of Pos_n as an open subset of a euclidean space. Just as with ordinary space, the tangent space at a point $p \in \text{Pos}_n$ is a translation of Sym_n. Without using more sophisticated language, we may identify it with Sym_n, just as we identify $\mathbf{R}^n$ with the tangent space at any point of $\mathbf{R}^n$.

The exponential map from Sym_n to Pos_n

We shall now relate Sym_n and Pos_n by means of the ordinary exponential map

$$\exp: \text{Mat}_n \to \text{GL}_n$$

given by the usual power series

$$\exp(v) = \sum_{k=0}^{\infty} \frac{v^k}{k!}.$$

If v, w commute (that is $vw = wv$), then $\exp(v + w) = \exp(v) \exp(w)$. Thus $\exp(v)$ is invertible, with inverse equal to $\exp(-v)$.

We shall use the exponential map to "bend" Sym_n. More precisely, we consider the exponential map just on the symmetric matrices

$$\exp: \text{Sym}_n \to \text{Pos}_n,$$

because the image of Sym_n under the exponential map lies in Pos_n. To see this, note that if v is symmetric, and we put $q = \exp(v/2)$, then q is symmetric and $\exp(v) = q^2$. Hence $\exp(v)$ is positive definite, because for all vectors $\xi \in \mathbf{R}^n, \xi \neq 0$, we have

$$\langle q^2\xi, \xi \rangle = \langle q\xi, q\xi \rangle > 0.$$

From linear algebra, the exponential map is a differential (i.e. C^∞) isomorphism, namely it has a C^∞ inverse, which can be called the **logarithm**. To see this, let p be a positive matrix. We can diagonalize p, that is there exists a basis $\{\xi_1, \ldots, \xi_n\}$ of $\mathbf{R}^n$ and numbers $\lambda_1, \ldots, \lambda_n > 0$ such that

$$p\xi_i = \lambda_i \xi_i \qquad \text{for} \quad i = 1, \ldots, n.$$

Then one defines $\log p = v$ to be the linear map represented by the diagonal matrix

$$\begin{pmatrix} \log \lambda_1 & & 0 \\ & \ddots & \\ 0 & & \log \lambda_n \end{pmatrix}$$

with respect to the basis $\xi_1, \ldots, \xi_n$. Similarly, one can define a square root of p to be the linear map represented by the matrix

$$\begin{pmatrix} \lambda_1^{1/2} & & 0 \\ & \ddots & \\ 0 & & \lambda_n^{1/2} \end{pmatrix}$$

with respect to the basis $\xi_1, \ldots, \xi_n$.

A variable positive definite scalar product

We come to one of our goals, which is to define a positive definite scalar product on Sym_n viewed as the tangent space at a point $p \in \mathrm{Pos}_n$, in such a way that this product depends on p. We define this product to be

$$\langle v, w \rangle_p = \mathrm{tr}(p^{-1}vp^{-1}w) \qquad \text{so that} \qquad |v|_p^2 = |v|_{p,\mathrm{tr}}^2 = \mathrm{tr}((p^{-1}v)^2).$$

The positive definiteness comes from the fact that

$$\mathrm{tr}(p^{-1}vp^{-1}v) = \mathrm{tr}(p^{-1/2}vp^{-1/2}p^{-1/2}vp^{-1/2}) = \mathrm{tr}(w^2)$$
$$\text{with} \quad w = p^{-1/2}vp^{-1/2},$$

and $\mathrm{tr}(w^2) > 0$ if $v \neq 0$. If $t \mapsto p(t)$ is a curve in Pos_n, then the derivative of the length is given by

$$(ds/dt)^2 = \mathrm{tr}(p(t)^{-1}p'(t))^2 \qquad \text{abbreviated} \qquad ds^2 = \mathrm{tr}((p^{-1}\,dp)^2).$$

We call the above metric the **trace metric**. If $\alpha\colon [a, b] \to \mathrm{Pos}_n$ is a curve in Pos_n, then its length is defined by

$$L(\alpha) = \int_a^b |\alpha'(t)|_{\alpha(t),\mathrm{tr}}\, dt.$$

[*See the comments at the end of* §4.] Given two points p_1, $p_2 \in \mathrm{Pos}_n$, we define the **distance** $d(p_1, p_2)$ to be the inf (greatest lower bound) of the lengths of all curves α joining p_1 and p_2 in Pos_n. All curves are assumed piecewise C^1, i.e. with continuous derivatives. It is an exercise in elementary analysis to prove that this defines a distance function on Pos_n, namely it is ≥ 0, it is symmetric, and satisfies the triangle inequality. Thus Pos_n is a metric space.

We now come to the main theorem.

Theorem 2.1. *The exponential map is metric semi increasing.*

The proof will be given in the third section. Here we make more comments.

First, the metric increasing property can be seen infinitesimally. That is, for each $v \in \mathrm{Sym}_n$, let

$$\exp'(v)\colon \mathrm{Sym}_n \to \mathrm{Sym}_n$$

denote the derivative of the exponential map at the point v; thus

$$\exp'(v)w = \lim_{t \to 0} \frac{\exp(v + tw) - \exp(v)}{t}.$$

Then $\exp'(v)$ is a linear map, and the metric increasing property states that for all $w \in \mathrm{Sym}_n$ we have

$$|w|_{\mathrm{tr}} \leq |\exp'(v)w|_{\exp v}.$$

Hence the inverse map is metric semi decreasing. This holds for the derivative, but then it also holds for the distance function, which is the inf of the lengths of curves, and the length is defined in terms of the derivative of the curve.

Corollary 2.2. Pos_n *is a complete metric space.*

Proof. Let $\{p_n\}$ be a Cauchy sequence in Pos_n, $p_n = \exp(v_n)$ with $v_n \in \mathrm{Sym}_n$. By the metric semi decreasing property, $\{v_n\}$ converges to some $v \in \mathrm{Sym}_n$. By continuity, one gets at once that $\{p_n\}$ converges to $\exp(v)$.

One defines a line through the origin in Sym_n as usual. One picks a "vector" $v \in \mathrm{Sym}_n$, $v \neq 0$, and the line through the origin in the direction of v is the curve $t \mapsto tv$ ($t \in \mathbf{R}$). For such lines, one has a stronger version than Theorem 2.1, as follows:

Theorem 2.3. *The exponential map* $\exp \colon \mathrm{Sym}_n \to \mathrm{Pos}_n$ *is metric preserving on a line through the origin. If* $p \in \mathrm{Pos}_n$, $p = \exp v$, $v \in \mathrm{Sym}_n$, *then* $d(e, p) = |v|_{\mathrm{tr}}$.

Proof. Such a line has the form $t \mapsto tv$ with some $v \in \mathrm{Sym}_n$, $v \neq 0$. We need to prove

$$|v|_{\mathrm{tr}}^2 = |\exp'(tv)v|_{\exp tv}^2.$$

Note that

$$\frac{d}{dt} \exp(tv) = \exp'(tv)v$$

$$= \frac{d}{dt} \sum \frac{t^n v^n}{n!}$$

$$- \sum \frac{t^{n-1}}{(n-1)!} v^n$$

$$= \exp(tv)v.$$

Hence

$$|\exp'(tv)v|_{\exp tv}^2 = \mathrm{tr}(((\exp tv)^{-1}(\exp tv)v)^2)$$

$$= \mathrm{tr}(v^2)$$

$$= |v|_{\mathrm{tr}}^2,$$

which proves the first assertion.

As to the second, i.e. the global version of the theorem, it comes as an application of Theorem 2.1. Let $p = \exp v$. Note that $|v|_{\mathrm{tr}}$ is the length of the line segment between 0 and v. To see that $d(e, p) = |v|_{\mathrm{tr}}$, let α be a piecewise C^1 curve in Pos_n between e and p, of length L. Its image under $\exp^{-1}$ (i.e. under the log) is a curve between 0 and v. As we know, $|v_{\mathrm{tr}}|$ is the distance between 0 and v in Sym_n. By Theorem 2.1, $\log \colon \mathrm{Pos}_n \to \mathrm{Sym}_n$

is metric semi decreasing. It follows that

$$|v|_{\mathrm{tr}} = d(0, v) \leqq L.$$

Hence the image under exp of the line segment between 0 and v is a curve between e and p, of shortest length, which is the distance between e and p. This proves the global version of Theorem 2.3.

With the above two theorems, we can prove the semi parallelogram law "at the origin," namely:

Theorem 2.4. *Let* $v_1 \in \mathrm{Sym}_n$, $v_1 \neq 0$. *Let* $v_2 = -v_1$. *Let* $x_1 = \exp(v_1)$, $x_2 = \exp(v_2)$ *and* $z = \exp(0) = e$. *Then for any* $v \in \mathrm{Sym}$ *and* $x = \exp(v)$, *we have*

$$d(x_1, x_2)^2 + 4d(x, z)^2 \leqq 2d(x, x_1)^2 + 2d(x, x_2)^2.$$

Proof. In the vector space Sym, we have the exact parallelogram law

$$d(v_1, v_2)^2 + 4d(v, 0)^2 = 2d(v, v_1)^2 + 2d(v, v_2)^2,$$

where $d(v, w) = |v - w|_{\mathrm{tr}}$ for $v, w \in \mathrm{Sym}_n$. Under the exponential map, the distances on the left side are preserved by Theorem 2.3, and the distances on the right are expanded, so we get precisely the semi parallelogram law in Pos_n, thus proving the theorem.

In order to see that Pos_n is a Bruhat–Tits space, we still have to show that the above result at the origin is valid everywhere, in a sense to be made precise. This requires considering Pos_n as a homogeneous space, which we now discuss.

The group $G = \mathrm{GL}_n(\mathbf{R})$ of invertible matrices acts on Pos_n as follows. For $g \in G$ we define $[g]: \mathrm{Pos}_n \to \mathrm{Pos}_n$ by the formula

$$[g]p = gp{}^tg \qquad \text{for} \quad p \in \mathrm{Pos}_n.$$

As usual, ${}^tg$ denotes the transpose of g. We note that $[g]p$ is indeed positive, because for all $\xi \in \mathbf{R}^n$, we have for the scalar product $\langle \ , \ \rangle$ on $\mathbf{R}^n$:

$$\langle [g]p\xi, \xi \rangle = \langle gp{}^tg\xi, \xi \rangle = \langle p{}^tg\xi, {}^tg\xi \rangle > 0.$$

It is immediately verified that for $g, g_1, g_2 \in G$ we have

$$[g^{-1}] = [g]^{-1} \qquad \text{and} \qquad [g_1 g_2] = [g_1][g_2].$$

Since, as we have remarked earlier, every positive matrix is the square of a positive matrix, it follows that G acts transitively on Pos_n. This means that given $p_1, p_2 \in \text{Pos}_n$ there exists $g \in G$ such that $[g]p_1 = p_2$. Indeed, if $p = q^2$ with q positive, then $p = [q]e$. If $p_1 = [q_1]e$ and $p_2 = [q_2]e$, then $p_2 = [q_2 q_1^{-1}]p_1$.

Theorem 2.5. *The association $g \mapsto [g]$ is a representation of G in the group of isometries of Pos_n, that is each $[g]$ is an isometry.*

Proof. First we note that $[g]$ can also be viewed as a map on the whole vector space Sym_n, and this map is linear as a function of such matrices. Hence its derivative is given by

$$[g]'(p)w = g w^t g \qquad \text{for all } w \in \text{Sym}_n.$$

Now we verify that $[g]$ preserves the scalar product, or the norm. We have:

$$
\begin{aligned}
|[g]'(p)w|^2_{[g]p} &= \text{tr}(([g]p)^{-1} g w^t g)^2) \\
&= \text{tr}((g p^t g)^{-1} g w^t g)^2) \\
&= \text{tr}({}^t g^{-1} p^{-1} g^{-1} g w^t g^t g^{-1} p^{-1} g^{-1} g w^t g) \\
&= \text{tr}({}^t g^{-1} p^{-1} w p^{-1} w^t g) \\
&= \text{tr}((p^{-1} w)^2) \\
&= |w|^2_p
\end{aligned}
$$

which proves the theorem.

Theorem 2.6. *The map $p \mapsto p^{-1}$ is an isometry.*

Proof. This will be left as an exercise to the reader. It's just calculus in vector spaces.

Theorem 2.7. *The space Pos_n with the trace metric is a Bruhat–Tits space, i.e. it satisfies the semi parallelogram law.*

Proof. The proof amounts to translating Theorem 2.4 to arbitrary points. Since the elements of G act as isometries, it will suffice to prove that given $x_1, x_2 \in \text{Pos}_n$, there exists $g \in G$ such that $g(x_1) = \exp(v_1)$ and $g(x_2) = \exp(-v_1)$ for some $v_1 \in \text{Sym}_n$. But first, by transitivity, we can find h_1 such that $h_1(x_2) = e$ and $h_1(x_1) = p$ for some $p \in \text{Pos}_n$. Now $p^{-1/4} \in \text{Pos}_n$, and we let h_2 be the map

$$h_2 \colon x \mapsto p^{-1/4} x p^{-1/4},$$

so $h_2 = [p^{-1/4}]$. Then $h_2(e) = p^{-1/2}$, $h_2(p) = p^{1/2}$. Hence $h_2 \circ h_1 = g$ maps x_1 on $p^{1/2}$ and x_2 on $p^{-1/2}$. Since there is some v_1 such that $p = \exp(v_1)$, we have obtained the desired element $g \in G$. This concludes the proof.

So the only thing left to do concerning the semi parallelogram law is to prove Theorem 2.1, which we do in the next section. However, there is one more result which does not depend on this law and which is basic. Although we have emphasized the semi parallelogram law, we give this basic theorem here. It is an explicit formula giving the distance between two points p, q in Pos_n.

Theorem 2.8. *Let $a_1, \ldots, a_n$ be the roots of the polynomial $\det(tp - q)$. Then*

$$d(p, q) = \sum (\log a_i)^2.$$

Proof. The technique is similar to that of Theorem 2.7, using translations. First suppose $p = e$ (the unit matrix), and $q = d$ is diagonal with diagonal components $a_1, \ldots, a_n$. Let v be the diagonal matrix with components $\log a_i$ $(i = 1, \ldots, n)$. Then $d = \exp(v)$. By Theorem 2.3, we have

$$d_{\text{Pos}}(e, d) = d_{\text{Sym}}(0, v) = |v^2|_{\text{tr}} = \sum (\log a_i)^2,$$

thus proving the formula in this case. We reduce the general case to the above special case. We claim that there exists $g \in G$ such that $[g]p = e$ and $[g]q = d$ is diagonal. Indeed, we may first translate p to e with some element $h_1 \in G$, so without loss of generality we may assume $p = e$. There exists an orthonormal basis of $\mathbf{R}^n$ which diagonalizes q, so there exists a diagonal matrix d and a real unitary matrix k (preserving the euclidean norm on $\mathbf{R}^n$) such that $q = kdk^{-1} = kd\,{}^tk$ because ${}^tk = k^{-1}$. But $e = kk^{-1} = k\,{}^tk$, so taking $[k]q$ proves our claim. Finally from the equations $gp\,{}^tg = e$ and $gq\,{}^tg = d$ with diagonal d, we get $p = g^{-1}\,{}^tg^{-1}$ and $q = g^{-1}d\,{}^tg^{-1}$, so

$$\det(tp - q) = \det(tg^{-1}\,{}^tg^{-1} - g^{-1}d\,{}^tg^{-1})$$
$$= \det(g)^{-2} \det(te - d).$$

Thus the roots of $\det(tp - q)$ are the same as the roots of $\det(te - d)$, and $\text{dist}(p, q) = \text{dist}(e, d)$. This proves the theorem.

§3. The Metric Increasing Property of the Exponential Map

This section is devoted to the proof of Theorem 2.1, essentially following [Mo 53]. We abbreviate

$$\mathrm{Mat}_n(\mathbf{R}) = \mathcal{A}.$$

We write Sym instead of Sym_n.

Let $v, w \in \mathcal{A}$. We define

$$F_v(w) = \exp(-v/2) \cdot \exp'(v)w \cdot \exp(-v/2) = e^{-v/2}\exp'(v)we^{-v/2}.$$

Note that

$$\exp'(v)w = \frac{d}{dt}\exp(v + tw)\Big|_{t=0}.$$

Directly from the definitions, we get

$$(1) \qquad \exp'(v)w = \sum_{n=0}^{\infty}\frac{1}{n!}\sum_{r+s=n-1}v^r w v^s.$$

Since powers of an element commute with each other, we note that $\exp(-v/2)$ commutes with powers v^r, v^s.

Lemma 3.1. *The maps F_v and $\exp'(v)$ are symmetric with respect to the* tr-*scalar product on $\mathcal{A}$. If $v \in \mathrm{Sym}$, then F_v and $\exp'(v)$ map* Sym *into itself.*

Proof. A routine verification gives for $u, v, w \in \mathcal{A}$:

$$\mathrm{tr}(F_v(w)u) = \sum_{n=0}^{\infty}\frac{1}{n!}\sum_{r+s=n-1}\mathrm{tr}(\exp(-v/2)v^r w v^s \exp(-v/2)u)$$

$$= \sum_{n=0}^{\infty}\frac{1}{n!}\sum_{r+s=n-1}\mathrm{tr}(wv^s\exp(-v/2)u\exp(-v/2)v^r)$$

$$= \mathrm{tr}(wF_v(u))$$

because $\exp(-v/2)$ commutes with v^r and v^s. This concludes the proof that F_v is symmetric with respect to the tr-scalar product. If $v \in \mathrm{Sym}$, then formula (1) shows that F_v maps Sym into itself. The statements about $\exp'(v)$ follow the same pattern of proof.

We define $L_v\colon \mathcal{A} \to \mathcal{A}$ to be left multiplication, $L_v(w) = vw$, and R_v is right multiplication. We let $D_v = L_v - R_v$, so

$$D_v(w) = vw - wv = [v, w] \qquad \text{for all } v, w \in \mathcal{A}.$$

Lemma 3.2. *Let $v \in$ Sym. Then D_v^2 is symmetric on* Sym. *In other words, for all $w, u \in$ Sym, we have*

$$\langle u, D_v^2 w \rangle_{\mathrm{tr}} = \langle D_v^2 u, w \rangle_{\mathrm{tr}},$$

or written in another way,

$$\mathrm{tr}(u D_v^2 w) = \mathrm{tr}(w D_v^2 u).$$

Proof. Again this is routine, namely:

$$\begin{aligned}
D_v(w) &= vw - wv, \\
D_v^2(w) &= v^2 w - 2vwv + wv^2, \\
(D_v^2 w)u &= v^2 wu - 2vwvu + wv^2 u, \\
w D_v^2 u &= wv^2 u - 2wvuv + wuv^2.
\end{aligned}$$

Applying tr to these last two expressions and using its basic property $\mathrm{tr}(xy) = \mathrm{tr}(yx)$ yields the proof of the lemma.

We recall that a symmetric operator B on a vector space with a positive definite scalar product is called **semipositive**, written $B \geq O$, if we have

$$\langle Bw, w \rangle \geq 0 \qquad \text{for all } w \text{ in the space.}$$

Then one defines $B_1 \geq B_2$ if $B_1 - B_2 \geq O$.

In the proofs that follow, we shall use two basic properties.

Spectral property. *Let M be a symmetric linear map of a finite dimensional vector space over $\mathbf{R}$, with a positive definite scalar product. Let $b \geq 0$. Let $f_0(t)$ be a convergent power series such that $f_0(t) \geq b$ for all t in an interval containing the eigenvalues of M. Then $f_0(M) \geq bI$.*

Proof. Immediate by diagonalizing the linear map with respect to a basis.

We also have:

Schwarzian property. *For all v, $w \in$ Sym,*

$$\mathrm{tr}((vw)^2) \leqq \mathrm{tr}(v^2 w^2).$$

For the convenience of the reader, we recall the proof. The matrices (linear maps) v and w can be simultaneously diagonalized, if one of them is positive definite, and in that case the inequality amounts to the usual Schwartz inequality. If both matrices are singular, then one can consider a matrix $w + \epsilon e$ with the identity matrix e, and $\epsilon > 0$. Then $w + \epsilon e$ is non-singular for all sufficiently small $\epsilon \neq 0$, and one can then use the preceding non-singular case, followed by taking a limit as $\epsilon \to 0$. This concludes the proof.

We define a formal power series

$$f(t) = \sum_{k=0}^{\infty} \frac{(t/2)^{2k}}{(2k+1)!} = \frac{\sinh t/2}{t/2} = \frac{\exp(t/2) - \exp(-t/2)}{t}.$$

We note that L_v and R_v commute with each other, and so

$$\exp(D_v/2) = \exp(L_v/2)\exp(-R_v/2).$$

We may take $f(D_v)$. Since only even powers of D_v occur in the power series for f, it follows that if $v \in$ Sym, then $f(D_v)$ maps Sym into itself, and the operator

$$f(D_v)\colon \mathrm{Sym} \to \mathrm{Sym}$$

is symmetric for the tr-scalar product.

Lemma 3.3. *For any $v \in \mathcal{A}$, we have $D_v F_v = D_v f(D_v)$.*

Proof. Let $t \mapsto x(t)$ be a smooth curve in $\mathcal{A}$. Then

$$x(\exp x) = (\exp x)x.$$

Differentiating both sides gives

$$x' \exp x + x(\exp x)' = (\exp x)'x + (\exp x)x',$$

and therefore

$$x' \exp x - (\exp x)x' = (\exp x)'x - x(\exp x)'.$$

Multiplying on the left and right by $\exp(-x/2)$, and using the fact that x commutes with $\exp(-x/2)$ yields

(2)
$$\exp(-x/2)x'\exp(x/2) - \exp(x/2)x'\exp(-x/2)$$
$$= \exp(-x/2)(\exp x)'\exp(-x/2)x$$
$$-x\exp(-x/2)(\exp x)'\exp(-x/2).$$

Since L_x and R_x commute, we have

$$\exp(D_x/2) = \exp(L_x/2)\exp(-R_x/2),$$

so (2) can be written in the form

(3)
$$(\exp(D_x/2) - \exp(-D_x/2))x' = D_x F_x x'.$$

We now take the curve $x(t) = v + tw$, and evaluate the preceding identity at $t = 0$, so $x'(0) = w$, to conclude the proof of the lemma.

Theorem 3.4. *Let $v \in$ Sym. Then $F_v = f(D_v)$ on Sym. Hence for $w \in$ Sym, we have*

$$\exp'(v)w = \exp(v/2) \cdot f(D_v)w \cdot \exp(v/2).$$

Proof. Let $h_v = F_v - f(D_v)$. Then h_v: Sym $\to$ Sym is symmetric linear, and its image is contained in the subspace $E = \mathrm{Ker} D_v \cap$ Sym. Since Sym is finite dimensional, it is the direct sum of E and its orthogonal complement $E^\perp$ in Sym. Since h_v is symmetric, it maps $E^\perp$ into $E^\perp$, but h_v also maps $E^\perp$ into E, so $h_v = 0$ on $E^\perp$. In addition, E is the commutant of v in Sym, and hence $f(D_v) = \mathrm{id} = F_v$ on E, so $h_v = 0$ on E. Hence $h_v = 0$ on Sym, thus concluding the proof of the theorem.

Theorem 3.5. *Let $v \in$ Sym. Then D_v^2 is semipositive, and $f(D_v) \geqq I$.*

Proof. By Lemma 3.2, for $w \in$ Sym we have

$$\langle D_v^2 w, w \rangle_{\mathrm{tr}} = \mathrm{tr}(wv^2w - 2vwvw + v^2w^2)$$
$$= 2\,\mathrm{tr}(v^2w^2 - (vw)^2).$$

Thus the semipositivity of D_v^2 results from the Schwarzian property of tr. Now we can write

$$f(t) = f_0(t^2)$$

where $f_0(t)$ is the obvious power series, which has positive coefficients. Note that $f_0(t) \geq 1$ for all $t \geq 0$. Therefore by the spectral property of power series, it follows that

$$f(D_v) = f_0(D_v^2) \geq I.$$

This concludes the proof.

Theorem 3.6. *The exponential map* exp *is* tr-*norm semi increasing on* Sym, *that is for all* v, $w \in$ Sym, *putting* $p = \exp(v)$, *we have*

$$|w|_{\mathrm{tr}}^2 = \mathrm{tr}(w^2) \leq \mathrm{tr}((p^{-1}\exp'(v)w)^2) = |\exp'(v)w|_{p,\mathrm{tr}}^2.$$

Proof. The right side of the above inequality is equal to

$$\begin{aligned}
\mathrm{tr}((p^{-1}\exp'(v)w)^2) &= \mathrm{tr}((\exp(-v/2)\cdot\exp(v)w\cdot\exp(-v/2))^2) \\
&= \mathrm{tr}(F_v(w)^2) \\
&= |f(D_v)w|_{\mathrm{tr}}^2 \quad \text{by Theorem 2.4.}
\end{aligned}$$

Applying Theorem 3.5 now concludes the proof.

Corollary 3.7. *For each* $v \in$ Sym, *the maps*

$$F_v \text{ and } \exp'(v)\colon \mathrm{Sym} \to \mathrm{Sym}$$

are linear automorphisms.

Proof. Theorem 3.6 shows that Ker $\exp'(v) = 0$, and $\exp'(v)$ is a linear isomorphism. The statement for F_v then follows because F_v is composed of $\exp'(v)$ and multiplicative translations by invertible elements in Sym. This concludes the proof.

We note that Theorem 3.6 concludes the proof of Theorem 2.1.

§4. Historical Notes

The presentation of the above material essentially follows a path which is the reverse of the historical path. It took almost a century before certain ideas were given their full generality and simplicity.

Historically, things start at the end of the nineteenth century. Klingenberg [Kl 83/95] asserts that von Mangoldt essentially proved what is called today

the Cartan–Hadamard theorem for surfaces [vM 1881], 15 years before
Hadamard did so [Ha 1896]. Actually, von Mangoldt refers to previous papers by others before him, Hadamard refers to von Mangoldt, and
Cartan [Ca 28] refers to Hadamard (Cartan dealt with arbitrary Riemannian
manifolds). I am unable to read the original articles.

Helgason [He 62] gave a proof of Cartan's fixed point theorem following
Cartan's ideas, see the revised version [He 78], Chapter I, Theorem 13.5,
namely: On a Riemannian manifold of seminegative curvature, a compact
group of isometries has a fixed point. Cartan's immediate purpose was to
show that all maximal compact subgroups of a semisimple Lie group are
conjugate. Mostow [Mo 53] gave a similar exposition, but in a more limited
context. They use a center of mass rather than the circumcenter.

Then Bruhat–Tits [BrT 72] formulated their fixed point theorem, here
stated as Theorem 1.2, setting up the parallelogram law condition prominently. Serre used a variation of their proof and the formulation of Theorem 1.1 to reach the currently ultimate result with the very simple proof
we have given. I don't know that Serre published this, but it is referred to
exactly as we have stated it in Brown [Bro 89], Chapter VI, Theorem 2 of
§5. Thus a line of thoughts which started a century before, abuts to a basic
elementary theorem about metric spaces. The condition of compactness
is replaced by the condition of boundedness, and the more complicated
notion of curvature is replaced by the semi parallelogram law.

In addition, the center of mass which occurred in Cartan's treatment
(and others following him), is replaced by the circumcenter, following
Bruhat–Tits.

Bruhat–Tits actually characterized simply connected Riemannian manifolds with seminegative curvature by the semi parallelogram law [BrT
72]. From that point on, a theory of curvature for metric spaces rather than
manifolds developed separately, with an extensive exposition in Ballman
[Ba 95], containing Theorem 1.1. Ballman refers to Brown for Theorem 1.1,
cf. [Ba 95], Theorem 5.1 and Proposition 5.10 of Chapter I.

The development of the theory of symmetric spaces is due to E. Cartan
[Ca 28], [Ca 46], [Ca 51]. Mostow gave a very elegant exposition of some
of Cartan's results in [Mo 53], and the proof of the present section §3 is
essentially taken from Mostow's exposition.

In standard basic analysis courses, one defines the **euclidean length** of
a curve in an open set U of $\mathbf{R}^N$ to be

$$L(\alpha) = \int_a^b |\alpha'(t)|\, dt,$$

where the norm is the standard euclidean norm. In the latter part of the
nineteenth century, various people were looking for geometries which were
different from the euclidean one, and variable metrics came under consideration, when the metric at each point depends on the point. A simple

example of a variable norm may have already been encountered in some courses.

For instance, let $\mathbf{D}$ be the unit disc in $\mathbf{R}^2$. For each point $x \in \mathbf{D}$ let $r = |x|$ be the euclidean distance from the origin. Given a vector $v \in \mathbf{R}^2$, one defines the **Lobatchevsky** or **Poincaré metric** by

$$|v|_x = \frac{1}{1 - r^2}|v|_{\text{euc}}, \qquad \text{with} \quad r = |x| = \sqrt{x_1^2 + x_2^2}.$$

Thus $|v|_x$ depends on the distance from the origin, and $|v|_x \to \infty$ as $|x| \to 1$. The dependence on x in the unit disc is still rather simple compared to the trace metric on the euclidean space Sym.

For a systematic account of how the exponential metric increasing property is related to the semi parallelogram law, see [La 99], where matters are discussed in the context of differential geometry. Note that from this context, one can also deduce that the disc with the Lobatchevsky–Poincaré metric is a Bruhat–Tits space, and in particular satisfies the semi parallelogram law. To see that this example is not essentially different from the one given with the exponential map needs a general theory of exponential maps, covering both situations which a priori do not look alike.

Bibliography

[Ba 95]   W. BALLMAN, *Lectures on Spaces of Nonpositive Curvature*, Birkhäuser, 1995

[BGS 85]   W. BALLMAN, M. GROMOV, and V. SCHROEDER, *Manifolds of Nonpositive Curvature*, Birkhäuser, 1985

[Bro 89]   K. BROWN, *Buildings,* Springer-Verlag, 1989

[BrT 72]   F. BRUHAT and J. TITS, Groupes Réductifs sur un Corps Local I, *Pub. IHES* **41** (1972) pp. 5–251

[Ca 27a]   E. CARTAN, Sur une classe remarquable d'espaces de Riemann, *Bull. Soc. Math. France* **54** (1927) pp. 114–134

[Ca 27b]   E. CARTAN, Sur certaines formes Riemanniennes remarquables des géometries à groupe fondamental simple, *Ann. Sci. Ecole Norm. Sup.* **44** (1927) pp. 345–467

[Ca 28/46] E. CARTAN, *Leçons sur la Géometrie des Espaces de Riemann*, Gauthiers-Villars, 1928; Second edition 1946

[Ca 51]   E. CARTAN, *Leçons sur la Géometrie des Espaces de Riemann II*, Gauthiers-Villars, 1951

[Ha 86]   J. HADAMARD, Les surfaces à courbures opposées et leur lignes géodesiques, *J. Math. Pures Appl.* **(5) 4** (1896) pp. 27–73

[He 62]   S. HELGASON, *Differential Geometry and Symmetric Spaces,* Academic Press, 1962

[He 78]   S. HELGASON, *Differential Geometry, Lie Groups, and Symmetric Spaces*, Academic Press, 1978

[K1 83/95]   W. KLINGENBERG, *Riemannian Geometry*, de Gruyter 1983; Second edition 1995

[La 99]   S. LANG, *Fundamentals of Differential Geometry*, Springer-Verlag 1999

[vM 81]   H. VON MANGOLDT, Über diejenigen Punkte auf positv gekrümmten Flächen, welche die Eigenschaft haben, dass die von ihnen ausgehenden geodätischen Linien nie aufhören, kurzeste Linien zu sein, *J. reine angew. Math.* **91** (1881) pp. 23–52

[Mo 53]   D. MOSTOW, *Some New Decomposition Theorems for Semisimple Groups,* Memoirs AMS, 1953

Harmonic and Symmetric Polynomials

We revert to a more algebraic topic. Courses in linear algebra or more general algebra at the undergraduate level provide foundations, but most often do not have the time to point to more sophisticated ways of putting the material together. The present topic illustrates concepts from linear algebra in a context which up to now has been reserved for much more advanced students. There is no difficulty in any given item, but just as the space of positive definite matrices provided a fertile ground to expand material from advanced calculus, the ring of polynomials and two important subsets, the harmonic polynomials and the symmetric polynomials, provide a fertile ground to illustrate concepts from linear algebra, not to speak of the calculus involved with partial derivatives. When dealing with polynomials, however, the analysis side of things is mostly illusory. Differentiation of polynomials is a purely algebraic notion, whereby we can define the derivative of a polynomial in one variable x to be

$$\frac{d}{dx}\sum_{k=0}^{d}x^k = \sum_{k=1}^{d}kx^{k-1}.$$

Similarly, for a polynomial in several variables, the partial derivatives can be defined by the analogous purely algebraic formula. No limits will be needed for the main theorems of this part.

Harmonic and Symmetric Polynomials

We are going to deal with polynomials in several variables over the real numbers. Some basic facts will be assumed, but will be summarized. We let the variables be $x_1, \ldots, x_n$. The set of all polynomials in these variables, with real coefficients will be denoted by

$$\mathbf{R}[x_1, \ldots, x_n].$$

A polynomial of the form $x_1^{j_1} \cdots x_n^{j_n}$ will be called a **simple monomial**. Any polynomial is a linear combination

$$P(x_1, \ldots, x_n) = \sum c_{j_1, \ldots, j_n} x_1^{j_1} \cdots x_n^{j_n},$$

with real coefficients $c_{(j)} \in \mathbf{R}$. We write a simple monomial in abbreviated form

$$M_{(j)}(x) = x^{(j)} = x_1^{j_1} \cdots x_n^{j_n},$$

so $P(x) = \sum c_{(j)} M_{(j)}(x)$. The **degree** of $M_{(j)}(x)$ is defined to be

$$\deg M_{(j)} = j_1 + \cdots + j_n.$$

The distinct monomials $M_{(j)}$ of given degree d are linearly independent over $\mathbf{R}$. A polynomial $P(x)$ as above is called **homogeneous** if all the monomials occurring with non-zero coefficient have the same degree d. For example, the polynomial

$$3x_1^3 x_2 - 7 \cdot x_1 x_2 x_3^2$$

is homogeneous of degree 4, but $3x_1^3 x_2 - x_1^2 x_2$ is not homogeneous. We denote by $\mathrm{Pol}(n, d)$ the vector space over $\mathbf{R}$ generated by the monomials of degree d in the n variables $x_1, \ldots, x_n$. Its dimension is equal to the number of simple monomials $M_{(j)}(x)$ of degree d. We won't use the fact, but it's not difficult to prove, that this number is the binomial coefficient

$$\binom{n - 1 + d}{n - 1}.$$

In other words,

$$\dim \mathrm{Pol}(n, d) = \binom{n - 1 + d}{n - 1}.$$

We can take derivatives of polynomials with the partial derivatives which we denote by $\partial_1, \ldots, \partial_n$. Putting the variable into the notation, these partial derivatives are the usual ones,

$$\partial_i = \frac{\partial}{\partial x_i} \qquad \text{for} \quad i = 1, \ldots, n.$$

The partial derivative of a polynomial is another polynomial. If $P(x)$ is homogeneous of degree $d \geq 1$, then $\partial_i P$ is a homogeneous polynomial of degree $d - 1$, or 0. Of course, if P has degree 0, then P is constant, and $\partial_i P = 0$ for each i. Note that we can form the differential operator

$$P(\partial_1, \ldots, \partial_n), \quad \text{abbreviated } P(\partial),$$

that is we can substitute the partial derivatives for the variables. For any polynomial Q, if $P = \sum c_{(j)} M_{(j)}$, then

$$P(\partial_1, \ldots, \partial_n) Q = \sum c_{(j)} \partial_1^{j_1} \cdots \partial_n^{j_n} Q.$$

§1. A Positive Definite Scalar Product

We define an example of a vector space with a scalar product, using Pol(n, d) and the partial derivatives as follows. For two polynomials

$$P, Q \in \text{Pol}(n, d)$$

we define the **scalar product**

$$\langle P, Q \rangle = (P(\partial)Q)(0).$$

Note that the bilinearity is immediate, that is

$$\langle P_1 + P_2, Q \rangle = \langle P_1, Q \rangle + \langle P_2, Q \rangle$$
$$\text{and} \quad \langle P, Q_1 + Q_2 \rangle = \langle p, Q_1 \rangle + \langle P, Q_2 \rangle.$$

If $c \in \mathbf{R}$, then $\langle cP, Q \rangle = c\langle P, Q \rangle = \langle P, cQ \rangle$.

Already more interesting is the fact that the scalar product is symmetric, that is $\langle P, Q \rangle = \langle Q, P \rangle$ for all homogeneous polynomials $P, Q \in \text{Pol}(n, d)$. To verify this property, it suffices to do it when P, Q are simple monomials. In this case, we have an explicit value for the scalar product.

Let $P = M_{(j)}(x)$ and $Q = M_{(k)}(x)$ be of degree d. If $(j) = (k)$, then

$$\langle M_{(j)}, M_{(j)} \rangle = j_1! \ldots j_n!.$$

If $(j) \neq (k)$, then

$$\langle M_{(j)}, M_{(k)} \rangle = 0.$$

In particular, the monomials form an orthogonal basis for Pol(n, d).

Proof. Suppose $(j) = (k)$, which means that $j_1 = k_1, \ldots, j_n = k_n$. Then

$$M_{(j)}(\partial) M_{(k)} = \partial_1^{j_1} \cdots \partial_n^{j_n} x_1^{j_1} \cdots x_n^{j_n}.$$

Taking the derivatives in the usual way, we find the stated value $j_1! \ldots j_n!$. Suppose $(j) \neq k$. Then for some i, $j_i \neq k_i$. Say $j_1 \neq k_1$. Consider

$$\partial_1^{j_1} x_1^{k_1}.$$

If $j_1 > k_1$, then we take more derivatives than the degree of x_1, and so we get 0. If $j_1 < k_1$, then $\partial_1^{j_1} x_1^{k_1}$ is equal to a constant times $x_1^{k_1 - j_1}$, so a positive power of x_1 is left. Evaluating at $(x) = 0$, so $x_1 = 0$, we get the value 0. This proves the orthogonality of the monomials, and also shows in particular that

$$\langle M_{(j)}, M_{(j)} \rangle > 0.$$

We now conclude that our scalar product is symmetric, because it is symmetric when we take the scalar product of two monomials. The symmetry in general follows from the bilinearity.

We can also conclude the positive definiteness, also from the bilinearity, and the fact that for $c_{(j)}, c_{(k)} \in \mathbf{R}$ we have

$$\langle c_{(j)} M_{(j)}, c_{(k)} M_{(k)} \rangle = \begin{cases} 0 & \text{if } (j) \neq (k), \\ c_{(j)}^2 j_1! \ldots j_n! & \text{if } (j) = (k). \end{cases}$$

Thus if $P = \sum c_{(j)} M_{(j)} \neq 0$, we get

$$\langle P, P \rangle = \sum c_{(j)}^2 j_1! \ldots j_n! > 0.$$

So we do have quite an interesting example of a positive definite scalar product, mixing polynomial algebra, and differentiation.

We mention one more formula. As before, we let $P \in \text{Pol}(n, d)$. Let $0 \leq m \leq d$.

Adjoint formula. *Let $R \in \text{Pol}(n, m)$ and $Q \in \text{Pol}(n, d - m)$. Then*

$$\langle R(\partial)P, Q \rangle = \langle P, RQ \rangle.$$

Proof. One line:

$$\langle P, RQ \rangle = (RQ)(\partial)P(0) = Q(\partial)R(\partial)P(0) = \langle R(\partial)P, Q \rangle,$$

because taking partial derivatives is a commutative operation. This proves the formula.

We shall study the set of all polynomials $\mathbf{R}[x_1, \ldots, x_n]$, which we denote by S, so

$$S = \mathbf{R}[x_1, \ldots, x_n].$$

We abbreviate

$$S^{(d)} = \text{Pol}(n, d)$$

for the subspace of homogeneous elements of degree d (other than 0).

At this point, we can take two directions which are independent. We can look at harmonic polynomials, which will be defined, or at symmetric polynomials together with a generalization of harmonic polynomials. We separate the two items, but you might want to read them simultaneously.

§2. Harmonic Polynomials

Let

$$\Delta = \partial_1^2 + \cdots + \partial_n^2.$$

From calculus of several variables, you should know that Δ is called the **Laplace operator**. A function f such that $\Delta f = 0$ is called **harmonic**. We are interested in harmonic polynomials, i.e. polynomials P such that $\Delta P = 0$. Note that

$$\Delta: \mathrm{Pol}(n, d) \to \mathrm{Pol}(n, d - 2)$$

is a linear map from the vector space of polynomials of degree d into the vector space of polynomials of degree $d - 2$.

Theorem 1. *This linear map is surjective. In other words, given a homogeneous polynomial Q of degree $d - 2$, there exists $P \in \mathrm{Pol}(n, d)$ such that $\Delta P = Q$.*

The theorem can be proved by brute force. We shall finesse another proof, by using more linear algebra.

We define $\mathrm{Har}(n, d)$ to be the set of all harmonic polynomials, homogeneous of degree d, or 0. It is clear that $\mathrm{Har}(n, d)$ is a subspace of $\mathrm{Pol}(n, d)$. Thus in the terminology of linear algebra,

$$\mathrm{Har}(n, d) = \mathrm{Ker}\,\Delta.$$

In other words, $\mathrm{Har}(n, d)$ is the kernel of the linear map

$$\Delta: \mathrm{Pol}(n, d) \to \mathrm{Pol}(n, d - 2).$$

We let

$$r^2 = r^2(x_1, \ldots, x_n) = x_1^2 + \cdots + x_n^2.$$

Thus r^2 is a homogeneous polynomial of degree 2. Observe that

$$r^2(\partial) = \Delta.$$

We have proved in general the transpose relation

$$\langle R(\partial)P, Q \rangle = \langle P, RQ \rangle.$$

We apply it to the special case $R = r^2$. Thus we find

$$\langle \Delta P, Q \rangle = \langle P, r^2 Q \rangle \qquad \text{for all } P \in \text{Pol}(n, d), \ Q \in \text{Pol}(n, d-2).$$

This puts us in a position to prove the surjectivity of Δ on $\text{Pol}(n, d-2)$. Suppose the image of Δ is not all of $\text{Pol}(n, d-2)$. Then there exists a polynomial $Q \in \text{Pol}(n, d-2)$, $Q \neq 0$, which is orthogonal to the image of Δ, by the linear algebra of positive definite scalar product. It follows that

$$0 = \langle \Delta P, Q \rangle = \langle P, r^2 Q \rangle$$

for all $P \in \text{Pol}(n, d)$. Hence $r^2 Q$ is orthogonal to all of $\text{Pol}(n, d)$. Since the scalar product is positive definite, we must have $r^2 Q = 0$, whence $Q = 0$, a contradiction which proves that

$$\Delta : \text{Pol}(n, d) \rightarrow \text{Pol}(n, d-2)$$

is surjective.

You are thus seeing linear algebra and positive definite scalar products at work in a context which is usually not covered in standard courses. We give one more example in the same vein.

Theorem 2. *The space of homogeneous polynomials of degree d admits an orthogonal decomposition*

$$\text{Pol}(n, d) = \text{Har}(n, d) + r^2 \text{Pol}(n, d-2).$$

This means that $\text{Pol}(n, d)$ is the sum of the two spaces on the right, and these two spaces $\text{Har}(n, d)$, $r^2 \text{Pol}(n, d-2)$ are orthogonal.

Proof. The general transpose relation shows that $\text{Har}(n, d)$ and $r^2 \text{Pol}(n, d-2)$ are orthogonal, because for $P \in \text{Har}(n, d)$ and $Q \in \text{Pol}(n, d-2)$ we have

$$\langle P, r^2 Q \rangle = \langle \Delta P, Q \rangle = 0$$

by the definition of P being harmonic. Hence

$$\dim[\text{Har}(n, d) + r^2 \text{Pol}(n, d-2)] = \dim \text{Har}(n, d) + \dim r^2 \text{Pol}(n, d-2).$$

But $\dim r^2 \text{Pol}(n, d-2) = \dim \text{Pol}(n, d-2)$ because multiplication by r^2 is an isomorphism

$$\text{Pol}(n, d-2) \overset{\approx}{\rightarrow} r^2 \text{Pol}(n, d-2).$$

Linear algebra also tells us that for any linear map on a vector space V, the dimension of the image plus the dimension of the kernel is equal to the dimension of the space. Here the vector space is $\text{Pol}(n, d)$, its kernel is the subspace of harmonic polynomials, and its image is $\text{Pol}(n, d - 2)$. Since

$$\text{Har}(n, d) + r^2 \text{Pol}(n, d - 2)$$

has the same dimension as $\text{Pol}(n, d)$, it must be equal to $\text{Pol}(n, d)$. This proves the theorem.

§3. Symmetric Polynomials

We let W be the group of permutations of the variables. If you wonder why we use the letter W, it's because in a fancy context Herman Weyl discovered something called the Weyl group. For the ordinary situation of ordinary polynomials, this Weyl group is just what we said above, the group of permutations of the variables. Nothing to be frightened of.

If σ is a permutation of $\{1, \dots, n\}$, and P is a polynomial, then we can define the polynomial σP to be the polynomial such that

$$(\sigma P)(x_1, \dots, x_n) = P(x_{\sigma(1)}, \dots, x_{\sigma(n)}).$$

So W also acts on polynomials. A polynomial P such that $\sigma P = P$ for all permutations σ is called **symmetric**. Note that the polynomial

$$r^2 = x_1^2 + \cdots + x_n^2$$

is symmetric, and so is the polynomial $x_1^d + \cdots + x_n^d$ for every positive integer d. We denote by S^W the space of all symmetric polynomials. If P, Q are symmetric, so is $P + Q$, PQ and for any scalar $c \in \mathbf{R}$, the polynomial cP is symmetric. Thus the set of symmetric polynomials is closed under addition, multiplication, and multiplication by scalars. Actually, S^W contains the constant polynomials.

We shall be interested in another kind of example of symmetric polynomials. Let T be a new variable, and consider the polynomial

$$F(T) = (T - x_1) \cdots (T - x_n).$$

If we expand it out, we get

$$F(T) = T^n - I_1 T^{n-1} + I_2 T^{n-2} - \cdots + (-1)^n I_n,$$

where $I_1, \ldots, I_n$ are polynomials in $x_1, \ldots, x_n$. For instance,

$$I_1 = I_1(x_1, \ldots, x_n) = x_1 + \cdots + x_n,$$
$$I_2 = I_2(x_1, \ldots, x_n) = x_1 x_2 + x_1 x_3 + \cdots + x_{n-1} x_n,$$
$$\cdots$$
$$I_n = I_n(x_1, \ldots, x_n) = x_1 \ldots x_n.$$

The specific, explicit determination of the polynomials $I_1, \ldots, I_n$ will not be needed. One can see a priori that they are symmetric in the variables $x_1, \ldots, x_n$ because the product

$$(T - x_1) \cdots (T - x_n)$$

is unchanged by a permutation of the factors $(T - x_i)$ $(i = 1, \ldots, n)$, so the coefficients of this polynomial are also unchanged. Observe that

$$\deg I_d = d.$$

One calls the polynomials $I_1, \ldots, I_d$ the **elementary symmetric polynomials**. A basic fact of algebra gives us:

Theorem 1. *The elementary symmetric polynomials generate all symmetric polynomials. More precisely, given a symmetric polynomial $P(x_1, \ldots, x_n)$ there exists a polynomial F such that*

$$P(x_1, \ldots, x_n) = F(I_1, \ldots, I_n).$$

Or, using another notation,

$$S^W = \mathbf{R}[I_1, \ldots, I_n].$$

The proof is not very difficult, and you can find it in some undergraduate texts, for instance [La 87], Chapter IV, Theorem 8.1. The proof is by induction on n, and also induction on d. Just for fun, carry out for yourself the case $n = 2$, that is show that $x_1 + x_2$ and $x_1 x_2$ generate all symmetric polynomials in two variables.

We want to see explicitly what is needed to get all polynomials from the symmetric ones. We use the terminology of linear algebra and vector spaces, even though S^W is not closed under taking quotients. Let $P_1, \ldots, P_N$ be elements of S. We say that $P_1, \ldots, P_N$ are **linearly independent** over S^W if given a relation

$$\sum_{j=1}^{N} F_j P_j = 0 \qquad \text{with } F_j \in S^W$$

we must have $F_j = 0$ for all j. We say that $\{P_1, \ldots, P_N\}$ is a **basis** of S over S^W if $P_1, \ldots, P_N$ are linearly independent over S^W, and if every element of S is a linear combination of $P_1, \ldots, P_N$ with coefficients in S^W; in other words, given $Q \in S$, there exist $F_1, \ldots, F_N \in S^W$ such that

$$Q = F_1 P_1 + \cdots + F_N P_N.$$

We claim that there exists a basis, and we want to exhibit such a basis explicitly. Actually we shall exhibit two types of bases, one more interesting than the other.

There is one very simple basis, which goes back to Kronecker in the 19th century. It is a simple exercise to show that the monomials

$$x_1^{r_1} \ldots x_n^{r_n} \qquad \text{with } 0 \leqq r_i \leqq n - i$$

form a basis. Note that there are exactly $n!$ such monomials. We won't give the proof. We want to get into a somewhat more complicated and more interesting basis, which connects with harmonic functions and has applications in several fields of mathematics, involving both algebra and analysis.

We let

$$S_+^W = \text{subspace of symmetric polynomials with zero constant term.}$$

We define a polynomial H to be **W-harmonic** if and only if

$$Q(\partial)H = 0 \qquad \text{for all } Q \in S_+^W.$$

Note that a plain harmonic polynomial has to satisfy only one equation, $\Delta P = 0$. Here, for a W-harmonic polynomial we are requiring an infinite number of equations, involving all $Q \in S_+^W$. One could get away with a finite number, namely the generators $I_1, \ldots, I_n$, but still, there is more than one equation if $n \geq 2$.

We let:

$$\text{Har}_W - \text{vector space of } W\text{-harmonic polynomials};$$

$$\text{Har}_W^{(d)} = \text{vector subspace of homogeneous elements of}$$
$$\text{degree } d \text{ in Har}_W.$$

Then we have an orthogonal sum decomposition

$$\text{Har}_W = \sum_{d=0}^{\infty} \text{Har}_W^{(d)}.$$

In other words, a polynomial is W-harmonic if and only if all its homogeneous components are W-harmonic, and these homogeneous components are mutually orthogonal. This is immediate from the fact that homogeneous polynomials of different degrees are orthogonal, as we have seen.

We shall denote by $S^W \operatorname{Har}_W$ the set of all polynomials of the form

$$\sum_{j=1}^{m} P_j H_j \qquad \text{with } P_j \in S^W \text{ and } H_j \in \operatorname{Har}_W.$$

The next result was proved by Chevalley in a much more general context than we are covering now [Ch 55].

Theorem 2. *Let $S = R[x_1, \ldots, x_n]$ be the polynomial ring over the reals, and S^W the subset of symmetric polynomials, namely those polynomials fixed under the permutations of the variables. Let Har_W be the vector space of W-harmonic polynomials. Then*

$$S = S^W \operatorname{Har}_W.$$

Furthermore, the space Har_W is finite dimensional over $\mathbf{R}$, of dimension

$$\dim_{\mathbf{R}} \operatorname{Har}_W = \#W.$$

If $\{H_1, \ldots, H_N\}$ ($N = \#W$) is a basis of Har_W over $\mathbf{R}$, then it is also a basis of S over S^W.

Thus we get a very clear picture of the way all polynomials are generated linearly in terms of the symmetric polynomials and the W-harmonic polynomials. The proof will take several steps.

We shall use the notation

$$S_+^W S = SS_+^W$$

to mean the set of all elements

$$\sum_{j=1}^{m} P_j Q_j \quad \text{with } P_j \in S_+^W \text{ and } Q_j \in S.$$

Note that $SS_+^W S = S_+^W SS = S_+^W S$.

We first show that we can decompose S into an orthogonal sum

(1) $$S = S_+^W S + \operatorname{Har}_W.$$

That the sum is orthogonal means that Har_W is the orthogonal complement of $S_+^W S$. Recall that if U is a subspace of a vector space V, then the orthogonal complement $U^\perp$ is the set of $v \in V$ such that $\langle u, v \rangle = 0$ for all $u \in U$. Thus (1) asserts that $\text{Har}_W = (S_+^W S)^\perp$. This is easily proved as follows. Let P be a polynomial orthogonal to $S_+^W S$. For $Q \in S_+^W$ we get

$$0 = \langle QS, P \rangle = \langle S, Q(\partial)P \rangle,$$

from which it follows that $Q(\partial)P = 0$, whence $P \in \text{Har}_W$. Thus

$$(S^W S)^\perp \subset \text{Har}_W.$$

The reverse inclusion follows by reversing the steps in this argument, so (1) is proved. Since homogeneous polynomials of different degrees are orthogonal, we get the homogeneous version for each positive integer d, namely an orthogonal sum decomposition

$$(1_d) \qquad S^{(d)} = \sum_{r=1}^{d} (S^W)^{(r)} S^{(d-r)} + \text{Har}_W^{(d)}.$$

We may now repeat inductively to get the first assertion of the theorem, namely

$$S = S^W \, \text{Har}_W.$$

Next we need two lemmas about $S_+^W S$, following Chevalley. We shall use standard notation. If $Q_1, \ldots, Q_m$ are elements of S^W, we define the **ideal** $(Q_1, \ldots, Q_m)$ of S^W to be the set of all linear combinations

$$P_1 Q_1 + \cdots + P_m Q_m \qquad \text{with } P_j \in S^W.$$

Lemma 1. *Let $Q_1, \ldots, Q_m \in S^W$ be such that Q_1 does not belong to the ideal $(Q_2, \ldots, Q_m)$ of S^W. Let $P_1, \ldots, P_m$ be homogeneous elements of S such that $\sum_{\nu=1}^{m} P_\nu Q_\nu = 0$. Then $P_1 \in S_+^W S$.*

Proof. The proof will use some basic algebraic facts. If you are not acquainted with them, then refer to the appendix of this section where they will be discussed and proved.

The first thing we use is the **projection operator** on W-invariants, namely the **operator** A (for **averaging**) defined by

$$A(P) = \frac{1}{\#W} \sum_{w \in W} w(P).$$

We apply this operator to the linear combination $\sum P_\nu Q_\nu = 0$. Since

$$w(PQ) = w(P)w(Q) \qquad \text{and} \qquad w(Q_\nu) = Q_\nu \quad \text{because} \quad Q_\nu \in S^W$$

we get

$$A(P_1)Q_1 + \cdots + A(P_m)Q_m = 0.$$

If $\deg P_1 = 0$, so P_1 is constant, then $P_1 = A(P_1)$, contradicting the hypothesis. Hence $\deg P_1 > 0$. Assume the lemma by induction for all relations

$$\sum_{\nu=1}^{m} R_\nu Q_\nu = 0$$

with homogeneous R_ν and $\deg R_1 < \deg P_1$. Let τ be a transposition of two variables, say x_j, x_k. Then $\tau(P_\nu) - P_\nu$ is divisible by $x_j - x_k$, that is

$$\tau(P_\nu) - P_\nu = (x_j - x_k)R_\nu$$

for some homogeneous polynomial R_ν of degree smaller than $\deg P_\nu$ (see the appendix), and we have the linear relation

$$R_1 Q_1 + \cdots + R_m Q_m = 0.$$

By induction, $R_1 \in S_+^W S$, or in other words, $\tau(P_1) \equiv P_1 \bmod S_+^W S$. Since W is generated by transpositions, it follows that

$$w(P_1) \equiv P_1 \bmod S_+^W S \qquad \text{for all } w \in W.$$

Hence $A(P_1) \equiv P_1 \bmod S_+^W S$. Since P_1 is homogeneous of strictly positive degree, the same is true for $A(P_1)$. Hence $A(P_1) \in S_+^W S$, whence finally $P_1 \in S_+^W S$, which proves the lemma.

Lemma 2. *Let $H_1, \ldots, H_m$ be homogeneous elements of S linearly independent mod $S_+^W S$ over the constants. Then $H_1, \ldots, H_m$ are linearly independent over S^W.*

Proof. Let $R_1 H_1 + \cdots + R_m H_m = 0$ be a relation of linear dependence with $R_i \in S^W$ not all 0. Without loss of generality, we may assume none of the coefficients $R_1, \ldots, R_m$ is 0. Furthermore we may also assume that the polynomials R_i are homogeneous, and $\deg R_i + \deg H_i = d$ is constant, independent of i.

Let $I_1, \ldots, I_n$ be the elementary symmetric polynomials. We have already mentioned that $S^W = R[I_1, \ldots, I_n]$. The monomials $I_1^{j_1} \ldots I_n^{j_n}$ are elements of S, and deg $I_1^{j_1} \ldots I_n^{j_n}$ refers to the degree in S. Let $Q_0, Q_1, \ldots$ be an ordering of these monomials by increasing degree, with $Q_0 = 1$. For each $i = 1, \ldots, m$ we have a linear expression

$$R_i = \sum_{\nu \geq 0} c_{i\nu} Q_\nu$$

with constant coefficients $c_{i\nu}$, and $c_{i\nu} = 0$ if deg $R_i \neq$ deg Q_ν. Collecting terms yields

$$0 = \sum_{i=1}^{m} R_i H_i = \sum_{\nu \geq 0} P_\nu Q_\nu \qquad \text{with} \qquad P_\nu = \sum_{i=1}^{m} c_{i\nu} H_i,$$

and P_ν is homogeneous of degree $d - $ deg Q_ν. We prove that $P_\nu = 0$ for all ν by induction. For $\nu = 0$, we note that $Q_0 \notin (Q_1, Q_2, \ldots)$, so by Lemma 1, $P_0 \in S_+^W S$, which contradicts the linear independence assumption on $H_1, \ldots, H_m$. Inductively, suppose that $P_0, \ldots, P_{r-1} = 0$. We can then argue in exactly the same way, because $Q_r \notin (Q_{r+1}, Q_{r+2}, \ldots)$. This concludes the proof of Lemma 2.

Lemma 3. *The* **R**-*vector space* Har$_W$ *is finite dimensional, and actually its dimension is* $\leq \#W$, *that is* $\leq n!$.

Proof. How long a proof one gives depends on how much algebra you want to use. If one wants to keep arguments as elementary as possible, then perhaps the fastest way is to argue as follows. We use the fact already mentioned that the $n!$ monomials

$$M_{(r)}(x) = x_1^{r_1} \ldots x_n^{r_n} \qquad \text{with} \quad 0 \leq r_i \leq n - i$$

form a basis of S over S^W. Just as with vector spaces over fields, it's routine to show that two bases of S over S^W have the same number of elements. Furthermore, if $H_1, \ldots, H_m$ are elements of S, linearly independent over S^W, then $m \leq n!$, that is m is at most the dimension of S over S^W.

In particular, let $H_1, \ldots, H_m$ be homogeneous elements of Har$_W$ linearly independent over **R**. By (1), they are linearly independent mod $S_+^W S$ over the constants, i.e. they satisfy the hypothesis of Lemma 2. By Lemma 2, they are linearly independent over S^W. By the preceding remarks, we conclude $m \leq \#W$, so Har$_W$ is finite dimensional over **R**, of dimension $\leq \#W$. This proves Lemma 3, but see the comments on field theory in the appendix to this section.

We can now conclude the proof of Theorem 2. Let $H_1, \ldots, H_m$ be a basis of Har_W over **R**. The first assertion of Theorem 2, which we have proved,

$$S = S^W \mathrm{Har}_W$$

shows that

$$S = S^W H_1 + \cdots + S_m^W,$$

and so $H_1, \ldots, H_m$ are linear generators of S over S^W. Lemma 2 shows that $H_1, \ldots, H_m$ are also linearly independent over S^W, so they form a basis of S over S^W. As already remarked in the proof of Lemma 3, we then have $m = \#W$. This concludes the proof of Theorem 2.

Appendix

In the preceding proof of Theorem 2, we have used some simple algebraic facts, which we are now going to go over, in case you have not encountered them previously.

The W-averaging projection. You should have seen projections in a course on linear algebra. Let V be a vector space. Let $A: V \to V$ be a linear map. Let U be a subspace of V. We say that A is a **projection** on U if the image of A is U, and if A leaves U elementwise fixed, that is $Au = u$ for all $u \in U$. If V has a positive definite scalar product and is finite dimensional, and if U is a subspace, then we always have the orthogonal projection. But in the application of Lemma 1, we are dealing with a different situation. The vector space is the set of all polynomials S. The subspace is S^W, where W is the group of permutations. Each w can be viewed as a linear automorphism of S, also satisfying the multiplicative property

$$w(PQ) = w(P)w(Q) \qquad \text{for all } P, Q \in S.$$

The linear maps of S into itself form a vector space. We define

$$A = \frac{1}{\#W} \sum_{w \in W} w.$$

Then $AQ = Q$ for all $Q \in W$, because $w(Q) = Q$, so the sum applied to Q gives $(\#W)Q$, and dividing by $\#W$ finally gives Q itself. Furthermore, for any $P \in S$, the element $A(P)$ is in S^W, that is $A(P)$ is a symmetric polynomial, because if $w_1 \in W$, then the map $w \to w_1 w$ is a permutation of W, so

$$w_1(A(P)) = \frac{1}{\#W} w_1 \sum_{w \in W} w(P) = \frac{1}{\#W} \sum_{w \in W} w_1 w(P) = A(P).$$

Thus A is a projection on S^W.

The divisibility property. In the proof of Lemma 1, we used the following fact:

Let $P \in S$ be a polynomial in the n variables $x_1, \ldots, x_n$, and let τ be a transposition of two variables, say τ interchanges x_j and x_k. Then $\tau(P) - P$ is divisible by $x_j - x_k$.

We give the complete proof, which is easy. It suffices to prove the assertion when P is a monomial. After renumbering the variables, we may assume without loss of generality that $j = 1$ and $k = 2$, so the monomial can be written in the form

$$P(x_1, \ldots, x_n) = x_1^{j_1} x_2^{j_2} Q(x_3, \ldots, x_n).$$

Since the transposition leaves $x_3, \ldots, x_n$ fixed, it now suffices to prove the statement for two variables, say x and y, That is, it suffices to prove:

Let x, y be two variables. Let j, k be integers ≥ 0. Then

$$x^j y^k - x^k y^j$$

is divisible by $x - y$.

Proof. If j or $k = 0$, then the usual binomial factorization shows that $x^j - y^j$ and $x^k - y^k$ are divisible by $x - y$. Suppose both $j, k \geq 1$. We can then factor

$$x^j y^k - x^k y^j = xy(x^{j-1} y^{k-1} - x^{k-1} y^{j-1}).$$

We then conclude the proof by induction.

Ideals. The third thing we used were ideals. Let J be a subset of S. In Lemma 1, we used $J = S_+^W S$. One says that J is an **ideal** if J is closed under addition, and multiplication by elements of S. Thus $S_+^W S$ is an ideal. Let $P, Q \in S$. We define P **congruent to** Q mod J, in symbols

$$P \equiv Q \mod J,$$

to mean that $P - Q \in J$. It is immediate that congruence mod J is an equivalence relation, that is for all P, Q, $R \in S$ we have:

$$P \equiv P \quad \mod J;$$
$$P \equiv Q \quad \text{implies} \quad Q \equiv P;$$
$$P \equiv Q \quad \text{and} \quad Q \equiv R \quad \text{implies} \quad P \equiv R.$$

Furthermore, if $P \equiv Q \mod J$, then $RP \equiv RQ \mod J$.

Let $H_1, \ldots, H_m \in S$. We say that $H_1, \ldots, H_m$ are **linearly independent mod J** to mean that a relation

$$c_1 H_1 + \cdots + c_m H_m \equiv 0 \quad \mod J \qquad \text{with } c_i \in \mathbf{R}$$

implies that $c_i = 0$ for all $i = 1, \ldots, m$.

Note that linear dependence has been used with respect to two types of coefficients: linear dependence over the scalars $\mathbf{R}$, and linear dependence of elements of S over S^W. The context makes clear which notion is meant. In Lemma 2 we showed that when $J = S_+^W S$, there is an implication from one of these conditions to the other.

Remarks on field theory. The use of the monomials $x_1^{r_1} \ldots x_n^{r_n}$ is really too special and ad hoc to be satisfactory. One can use arguments from field theory instead, applying to very general situations as follows. Every element P of S is the root of a polynomial of degree #W, with coefficients in S^W, namely the polynomial

$$F(T) = \prod_{w \in W} (T - wP).$$

Indeed, this polynomial contains $T - P$ as a factor (using $w =$ identity), so $F(P) = 0$. Let $c_1, \ldots, c_n$ be distinct scalar, and let

$$Y = c_1 x_1 + \cdots + c_n x_n.$$

Then all the elements wY (with $w \in W$) are distinct, and there are #W such elements. Let Rat $= \mathbf{R}(x_1, \ldots, x_n)$ be the field of rational functions in $x_1, \ldots, x_n$ (quotients of polynomials), and $K = \mathbf{R}(I_1, \ldots, I_N)$ the rational functions in $I_1, \ldots, I_N$. Then Rat is a vector space over K, and ordinary linear algebra can be applied. It is fairly simple to show that the dimension of Rat over K is $\leq$#W because of the polynomial equation $F(Y) = 0$ mentioned above. Further arguments of field theory show that this dimension must in fact be exactly #W. If you want to see all this carried out in detail, check out my *Undergraduate Algebra* [La 87], Chapter VII, Theorem 4.9

(Artin's theorem). The present situation with symmetric polynomials gives an example of that theorem.

§4. Eigenfunctions and Characters

In this section, we deal with differential operators and get examples of eigenfunctions for them. We continue with the variables $x_1, \ldots, x_n$ and with the partial derivatives

$$\partial_i = \frac{\partial}{\partial x_i} \qquad \text{for} \quad i = 1, \ldots, n.$$

If $P \in S = \mathbf{R}[x_1, \ldots, x_n]$, then as before, we can form the differential operator

$$P(\partial) = P(\partial_1, \ldots, \partial_n) \in \mathbf{R}[\partial_1, \ldots, \partial_n].$$

One sometimes calls $P(\partial)$ a **differential operator with constant coefficients**. We define a **W-harmonic function** to be an infinitely differentiable function f on $\mathbf{R}^n$ such that $P(\partial)f = 0$ for all $P \in S_+^W$.

Theorem 1. *The W-harmonic functions on $\mathbf{R}^n$ are just the polynomial W-harmonic functions, so just the elements of* Har_W.

Proof. Let $I_1, \ldots, I_n$ as before be the elementary symmetric polynomials of the variables $x_1, \ldots, x_n$. Then for each i, x_l is a root of the symmetric polynomial

$$0 = x_i^n - I_1 x_i^{n-1} + \cdots + (-1)^n I_n.$$

Hence $x_i^n \in S_+^W S$ for $i = 1, \ldots, n$, and therefore

$$\partial_i^n \in S(\partial) S_+^W(\partial),$$

so ∂_i^n annihilates every W-harmonic function. Thus finally, if f is W-harmonic, then $\partial_i^n f = 0$ for all i, whence f is a polynomial.

Let V be a vector space over $\mathbf{R}$. Let $L: V \to V$ be a linear map. By an **eigenvector** v of A we mean an element of V for which there exists a number $\lambda \in \mathbf{R}$ such that $L(v) = \lambda v$. In our applications, V will consist of functions, and so instead of saying eigenvector, one says **eigenfunction** for L.

Specifically, we consider S as a vector space over $\mathbf{R}$. Each polynomial differential operator $P(\partial)$ with $P \in S$ then is a linear map of S into itself.

We want to analyze the eigenfunctions (eigenpolynomials) for all such operators $P(\partial)$. By definition, we let $S(\partial)$ be the set of all polynomial differential operators $P(\partial)$ with $P \in S$. By an **eigenfunction** f for $S(\partial)$ we mean a non-zero function f such that for each $D \in S(\partial)$ there is a real number $\lambda(D)$ such that

$$Df = \lambda(D)f.$$

Observe that for such a function f, the map $D \to \lambda(D)$ is a linear map of $S(\partial)$ into $\mathbf{R}$, which also satisfies the multiplicative property

$$\lambda(D_1 D_2) = \lambda(D_1)\lambda(D_2).$$

Quite generally, a linear map $\lambda \colon S \to \mathbf{R}$ which satisfies the multiplication rule

$$\lambda(PQ) = \lambda(P)\lambda(Q) \qquad \text{for all } P, Q \in S$$

will be called a **character** of S. Similarly, a linear map

$$\lambda \colon S(\partial) \to \mathbf{R}$$

satisfying the multiplicative rule $\lambda(D_1 D_2) = \lambda(D_1)\lambda(D_2)$ will be called a **character of** $S(\partial)$.

Example of characters. For each n-tuple $s = (s_1, \ldots, s_n)$ of real numbers, the evaluation map

$$P \mapsto P(s) \qquad \text{or} \qquad P(x_1, \ldots, x_n) \mapsto P(s_1, \ldots, s_n)$$

is a character of S, which we shall denote by λ_s. There no other characters, because if $\lambda \colon S \to \mathbf{R}$ is a character, let $s_i = \lambda(x_i)$ for $i = 1, \ldots, n$. Then by the linearity and multiplicative property of characters, we must have

$$\lambda = \lambda_s.$$

Example of eigenfunctions. Again let $s = (s_1, \ldots, s_n)$ be an n-tuple of real numbers. We define the function

$$f_s \colon \mathbf{R}^n \to \mathbf{R} \qquad \text{by} \qquad f_s(x_1, \ldots, x_n) = e^{s_1 x_1} \cdots e^{s_n x_n}.$$

Then

$$(\partial_i f_s)(x) = s_i f_s(x) \qquad \text{or} \qquad \partial_i f_s = s_i f_s.$$

Hence for any polynomial $P \in S$, we get

$$(P(\partial) f_s)(x) = P(s) f_s(x).$$

Hence f_s is an eigenfunction of S, with eigencharacter λ_s.

It's a nice problem to determine all eigenfunctions of S^W. This problem was solved in a very general context by Steinberg [St 64]. We shall give the solution only in a special case, because the general case requires more extensive tools. But the basic idea for the proof is the same in the special case and the general case.

Let λ be a character of S. Let $w \in W$. Then we can define the action of w on λ, that is $w\lambda$, by the formula

$$(w\lambda)(P) = \lambda(w^{-1}P).$$

It follows that $w\lambda$ is also a character of S. If $\lambda = \lambda_s$, so λ corresponds to an n-tuple $s = (s_1, \ldots, s_n)$, then $w\lambda_s$ corresponds to a permutation of $(s_1, \ldots, s_n)$, actually corresponds to

$$\left(s_{w(1)}, \ldots, s_{w(n)} \right).$$

We shall need a special case of a theorem proved in great generality by E. Artin.

Let $\lambda_1, \ldots, \lambda_m$ be distinct non-zero characters of S. Then they are lin-early independent over **R**.

Proof. We give Artin's proof, valid in much greater generality. The proof is by induction on m. Suppose $m = 1$. Then there cannot be a relation

$$c_1\lambda_1 = 0, \qquad \text{with } c_1 \in \mathbf{R}, \ c_1 \neq 0,$$

otherwise let P be such that $\lambda_1(P) \neq 0$. Then $c_1\lambda_1(P) - 0$ implies

$$\lambda_1(P) = 0,$$

a contradiction. Now assume the theorem by induction for $m - 1$ distinct non-zero characters. Suppose we have a relation

(1) $$c_1\lambda_1 + \cdots + c_m\lambda_m = 0$$

with not all coefficients c_j being 0. If one coefficient is 0, then we get a relation between at most $m - 1$ distinct non-zero characters, which is impossible by induction. So $c_j \neq 0$ for $j = 1, \ldots, m$, and $m \geq 2$. By assumption, there is an element $P \in S$ such that $\lambda_1(P) \neq \lambda_2(P)$. We then use high school elimination to get a relation between less than m characters.

By (1), we have

$$c_1\lambda_1(PQ) + \cdots + c_m\lambda_m(PQ) = 0 \qquad \text{for all } Q \in S.$$

By the multiplicativity property, this yields

$$(2) \quad c_1\lambda_1(P)\lambda_1(Q) + c_2\lambda_2(P)\lambda_2(Q) + \cdots + c_m\lambda_m(P)\lambda_m(Q) = 0$$

for all $Q \in S$. We multiply (1) by $\lambda_1(P)$, and evaluate at Q, to get

$$(3) \quad c_1\lambda_1(P)\lambda_1(Q) + c_2\lambda_1(P)\lambda_2(Q) + \cdots + c_m\lambda_1(P)\lambda_m(Q) = 0.$$

Subtracting (3) from (2) eliminates the first term, and gives

$$(4) \quad c_2(\lambda_2(P) - \lambda_1(P))\lambda_2(Q) + \cdots + c_m(\lambda_m(P) - \lambda_1(P))\lambda_m(Q) = 0$$

for all $Q \in S$. Since $\lambda_2(P) - \lambda_1(P) \neq 0$, and $c_2 \neq 0$, we see that (4) gives a linear relation with non-zero coefficients between at most $m - 1$ distinct non-zero characters, which contradicts the induction hypothesis, and concludes the proof.

Let λ be a character of S. We let λ_W be its restriction to S^W. We let

$$\text{Ei}(S^W(\partial), \lambda_W)$$

be the set of analytic functions on $\mathbf{R}^n$ which are eigenfunctions of $S^W(\partial)$ with character λ_W, or 0. By an **analytic function** we mean a function whose Taylor series converges to the function. The eigenfunction property means that the function F satisfies

$$P(\partial)F = \lambda_W(P)F \qquad \text{for all} \quad P \in S^W.$$

Note that $\text{Ei}(S^W(\partial), \lambda_W)$ is a vector space over $\mathbf{R}$. It is a fact that such analytic eigenfunctions must be of the form Qf where Q is a polynomial, and f is a character, but we are going to reprove this fact another way. We call

$$\text{Ei}(S^W(\partial), \lambda_W)$$

the λ_W-**eigenspace** of $S^W(\partial)$. Steinberg determined this eigenspace in general [St 64]. We describe it and give a proof in the special case when the character λ is of the form λ_s with an n-tuple $s = (s_1, \ldots, s_n)$ such that the numbers $s_1, \ldots, s_n$ are distinct. In this case, the character $\lambda = \lambda_s$ is called **regular**.

Theorem 2. *Let λ be a regular character. Then*

$$\dim \text{Ei}(S^W(\partial), \lambda_W) = \#W.$$

Let $\lambda = \lambda_s$ and let $f = f_s$ be the exponential function

$$f(x) = e^{s_1 x_1} \cdots e^{s_n x_n}.$$

Then the functions wf with $w \in W$ form a basis for the eigenspace. In other words, the functions obtained from f by a permutation of the variables form a basis for the eigenspace.

Proof. Since the coordinates $s_1, \ldots, s_n$ are distinct, we conclude that the functions wf obtained from f by a permutation of the variables are distinct, and they are characters, so they are linearly independent by Artin's theorem. We shall now prove that the dimension of $\mathrm{Ei}(S^W(\partial), \lambda_W)$ is at most $\#W$. This will conclude the proof of Steinberg's theorem.

Let $\{H_1, \ldots, H_m\}$ be a basis for Har_W. Map the eigenspace into $\mathbf{R}^n$ by the mapping

$$f \mapsto ((H_1(\partial)f)(0), \ldots, (H_m(\partial)f)(0)).$$

The map is linear, and it suffices to prove that it is injective. If the image of f is $(0, \ldots, 0)$, then $(H(\partial)f)(0) = 0$ for all $H \in \mathrm{Har}_W$. By Theorem 2, $S = S^W \mathrm{Har}_W$. Let $P \in S^W$. Then for some scalar μ we get

$$H(\partial)P(\partial)f = \mu H(\partial)f,$$

so for all $Q \in S$ we get $(Q(\partial)f)(0) = 0$. Hence the Taylor series of f at 0 is identically zero. Since f is real analytic, it follows that $f = 0$, whence the injectivity and the bound $\#W$ on the dimension of the eigenspace. This concludes the proof of Theorem 2.

Remark. Steinberg's proof in general follows the same pattern, but instead of having just the exponential functions as eigenfunctions, one has to multiply them by certain polynomials which can be explicitly determined. Things get somewhat more technical, and this is a good time to quit.

Bibliography

[Ch 55]   C. CHEVALLEY, Invariants of finite groups generated by reflections, *Amer. J. Math.* **77** (1955) pp. 778–782

[La 87]   S. LANG, *Undergraduate Algebra,* Springer-Verlag, 1987

[La 93]   S. LANG, *Algebra,* Third edition, Addison Wesley, 1993

[St 64]   R. STEINBERG, Differential equations invariant under finite reflection groups, *Trans. Amer. Math. Soc.* **112** (1964) pp. 392–400